Dynamic Reconfigurable Architectures and Transparent Optimization Techniques

Antonio Carlos Schneider Beck Fl. • Luigi Carro

Dynamic Reconfigurable Architectures and Transparent Optimization Techniques

Automatic Acceleration of Software Execution

Springer

Prof. Antonio Carlos Schneider Beck Fl.
Instituto de Informática
Universidade Federal do Rio Grande
do Sul (UFRGS)
Caixa Postal 15064
Campus do Vale, Bloco IV
Porto Alegre
Brazil
caco@inf.ufrgs.br

Prof. Luigi Carro
Instituto de Informática
Universidade Federal do Rio Grande
do Sul (UFRGS)
Caixa Postal 15064
Campus do Vale, Bloco IV
Porto Alegre
Brazil
carro@inf.ufrgs.br

ISBN 978-94-007-9073-5 ISBN 978-90-481-3913-2 (eBook)
DOI 10.1007/978-90-481-3913-2
Springer Dordrecht Heidelberg London New York

Softcover re-print of the Hardcover 1st edition 2010

Printed on acid-free paper

Springer is part of Springer Science+Business Media (www.springer.com)

To Sabrina,
for her understanding and support
To Antônio and Léia,
for the continuous encouragement

To Ulisses, may his journey be full of joy
To Érika, for all our moments
To Cesare, Esther and Beti, for being there

Preface

As Moore's law is losing steam, one already sees the phenomenon of clock frequency reduction caused by the excessive power dissipation in general purpose processors. At the same time, embedded systems are getting more heterogeneous, characterized by a high diversity of computational models coexisting in a single device. Therefore, as innovative technologies that will completely or partially replace silicon are arising, new architectural alternatives are necessary.

Although reconfigurable computing has already shown to be a potential solution when it comes to accelerate specific code with a small power budget, significant speedups are achieved just in very dedicated dataflow oriented software, failing to capture the reality of nowadays complex heterogeneous systems. Moreover, one important characteristic of any new architecture is that it should be able to execute legacy code, since there has already been a large amount of investment into writing software for different applications. The wide spread usage of reconfigurable devices is still withheld by the need of special tools and compilers, which clearly preclude reuse of legacy code and its portability.

The authors have written this book with the aforementioned limitations in mind. Therefore, this book, which is divided in seven chapters, starts presenting the main challenges computer architectures are facing these days. Then, a detailed study on the usage of reconfigurable systems, their main principles, characteristics, potential and classifications is done. A separate chapter is dedicated to present several case studies, with a critical analysis on their main advantages and drawbacks, and the benchmarks used for their evaluation. This analysis will demonstrate that such architectures need to attack a diverse range of applications with very different behaviors, besides supporting code compatibility, that is, the need for no modification in the source or binary codes. This proves that more must be done to bring reconfigurable computing to be used as main stream computing: dynamic optimization techniques. Therefore, binary Translation and different types of reuse, with several examples, are evaluated. Finally, works that combine both reconfigurable systems and dynamic techniques are discussed, and a quantitative analysis of one of these examples is presented. The book ends with some directions that could inspire new fields of research.

The main purpose of this book is to introduce reconfigurable systems and dynamic optimization techniques to the readers, using several examples, so it can be a source of reference whenever the reader needs. The authors hope you enjoy it, as they have enjoyed making the research that resulted in this book.

Porto Alegre

Antonio Carlos Schneider Beck Fl.
Luigi Carro

Acknowledgements

The authors would like to express their gratitude to the friends and colleagues at Instituto de Informatica of Universidade Federal do Rio Grande do Sul, and to give a special thanks to all the people in the Embedded Systems laboratory, who during several moments contributed for this research for many years.

The authors would also like to thank the Brazilian research support agencies, CAPES and CNPq.

Contents

Acronyms

ADPCM	Adaptive Differential Pulse-Code Modulation
ALU	Arithmetic Logic Unit
AMIL	Average Merged Instructions Length
ASIC	Application-Specific Integrated Circuit
ASIP	Application-Specific Instruction Set Processor
ATR	Automatic Target Recognition
BB	Basic Block
BHB	Block History Buffer
BT	Binary Translator
CAD	Computer-Aided Design
CAM	Content Addressable Memory
CCA	Configurable Compute Accelerator
CCU	Custom Computing Unit
CDFG	Control Data Flow Graph
CISC	Complex Instruction Set Computer
CLB	Configurable Logic Block
CM	Configuration Manager
CMOS	Complementary MetalOxide Semiconductor
CMS	Code Morphing Software
CPII	Cycles Per Issue Interval
CPLD	Complex Programmable Logic Device
CRC	Cyclic Redundancy Check
DADG	Data Address Generator
DAISY	Dynamically Architected Instruction Set from Yorktown
DCT	Discrete Cosine Transformation
DES	Data Encryption Standard
DFG	Data Flow Graph
DIM	Dynamic Instruction Merging
DLL	Dynamic-Link Library
DSP	Digital Signal Processing
DTM	Dynamic Trace Memoization

FFT Fast Fourier Transform
FIFO First In, First OutFirst In, First Out
FIR Finite Impulse Response
FO4 Fanout-Of-Four
FPGA Field-Programmable Gate Array
FU Functional Unit
GCC GNU Compiler Collection
GPP General Purpose Processor
GSM Global System for Mobile Communications
HDL Hardware Description Language
I/O Input-Output
IC Integrated Circuit
IDCT Inverse Discrete Cosine Transform
IDEA International Data Encryption Algorithm
ILP Instruction Level Parallelism
IPC Instructions Per Cycle
IPII Instructions Per Issue Interval
IR Instruction Reuse
ISA Instruction Set Architecture
ITRS International Technology Roadmap for Semiconductors
JIT Just-In-Time
JPEG Joint Photographic Experts Group
LRU Least Recently Used
LUT Lookup Table
LVP Load Value Prediction
MAC multiplier-accumulator
MAC Multiply Accumulate
MC Motion Compensation
MIMD Multiple Instruction, Multiple Data
MIN Multistage Interconnection Network
MIR Merged Instructions Rate
MMX Multimedia Extensions
MP3 MPEG-1 Audio Layer 3
MPEG Moving Picture Experts Group
NMI Number of Merged Instructions
OFDM Orthogonal frequency-division multiplexing
OPI Operation per Instructions
OS Operating System
PAC Processing Array Cluster
PACT-XPP eXtreme Processing Plataform
PAE Processing Array Elements
PC Program Counter
PCM Pulse-Code Modulation
PDA Personal Digital Assistant
PE Processing Element

PFU	Programmable Functional Units
PRISM	Processor Reconfiguration through Instruction Set Metamorphosis
RAM	Random Access Memory
RAW	Read After Write
RAW	Reconfigurable Architecture Workstation
RB	Reuse Buffer
RC	Reconfigurable Cell
REMARC	Reconfigurable Multimedia Array Coprocessor
RFU	Reconfigurable Functional Unit
RISC	Reduced Instruction Set Computer
RISP	Reconfigurable Instruction Set Processor
ROM	Read Only Memory
RRA	Reconfigurable Arithmetic Array
RST	Reuse through Speculation on Traces
RT	Register Transfer
RTM	Reuse Trace Memory
RU	Reconfigurable Unit
SAD	Sum of Absolute Difference
SCM	Supervising Configuration Manager
SDRAM	Synchronous Dynamic Random Access Memory
SIMD	Single Instruction, Multiple Data
SMT	Simultaneous multithreading
SoC	System-On-a-Chip
SSE	Streaming SIMD Extensions
VHDL	VHSIC Hardware Description Language
VLIW	Very Long Instruction Word
VMM	Virtual Machine Monitor
VP	Value prediction
VPT	Value Prediction Table
WAR	Write After Read
WAW	Write After Write
XREG	Exchange Registers

Chapter 1
Introduction

Abstract This introductory chapter presents several challenges that architectures are facing these days, such as the imminent end of the Moore's law as it is known today; the usage of future technologies that will replace silicon; the stagnation of ILP increase in superscalar processors and their excessive power consumption and, most importantly, how the aforementioned aspects are impacting on the development of new architectural alternatives. All these aspects point to the fact that new architectural solutions are necessary. Then, the main reasons that motivated the writing of this book are shown. Several aspects are discussed, as the why ILP does not increase as before; the use of both combinational logic and reconfigurable fabric to speedup execution of data dependent instructions; the importance of maintaining binary compatibility, which is the possibility of reusing previously compiled code without any kind of modification; yield issues and the costs of fabrication. This chapter ends with a brief review of what will be seen in the rest of the book.

1.1 Challenges

The possibility of increasing the number of transistors inside an integrated circuit with the passing years, according to Moore's Law, has been pushing performance at the same level of growth. However, this law, as known today, will no longer exist in a near future. The reason is very simple: physical limits of silicon [7, 15]. Because of that, new technologies that will completely or partially replace silicon are arising. However, according to the ITRS roadmap [12] these technologies have a high level of density and are slower than traditional scaled CMOS, or the opposite: new devices can achieve higher speeds but with a huge area and power overhead, even when comparing to future CMOS technology.

Additionally, high performance architectures as the diffused superscalar machines are achieving their limits. According to what is discussed in [5] and [13], there are no novelties in such systems. The advances in ILP (Instruction Level Parallelism) exploitation are stagnating: considering the Intel's family of processors, the overall efficiency (comparison of processors performance running at the same

A.C. Schneider Beck Fl., L. Carro, *Dynamic Reconfigurable Architectures and Transparent Optimization Techniques*,
DOI 10.1007/978-90-481-3913-2_1,

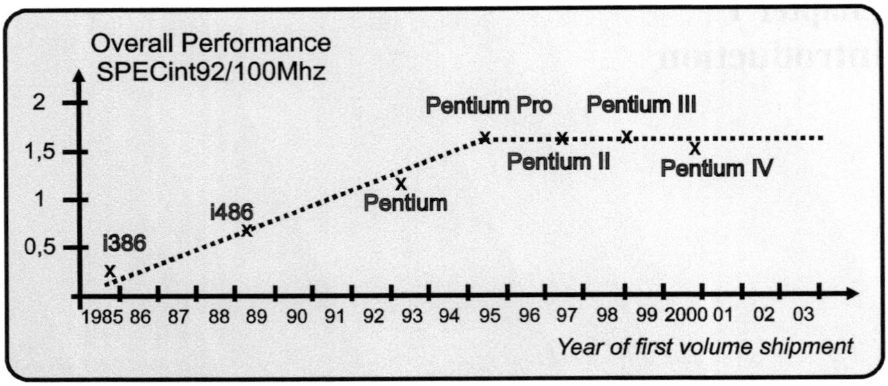

Fig. 1.1 There is no improvements regarding the overall performance in the Intel's Family of processors

clock frequency) has not significantly increased since the Pentium Pro in 1995, as Fig. 1.1 illustrates. The newest Intel architectures follow the same trend: the Core2 micro architecture has not presented a significant increase in its IPC (Instructions per Cycle) rate, as demonstrated in [10].

That is because these architectures are challenging some well-known limits of the ILP [19]. Therefore, the process of trying to increase the ILP has become extremely costly. In [3], a study on how the dispatch width affects the processor area is done. For instance, considering a typical superscalar processor based on the MIPS R10000, the register bank area grows cubically with the dispatch width. Consequently, recent increases in performance have occurred mainly thanks to boosts in clock frequency, through the employment of deeper pipelines. Even this approach, though, is reaching its limit.

In [1], the so-called "Mobile Supercomputers" are discussed. In the future, embedded devices will need to perform some intensive computational programs, such as real-time speech recognition, cryptography, augmented reality etc, besides the conventional ones, like word and e-mail processing. Figure 1.2 shows that even considering desktop computer processors, new architectures may not meet the requirements for future embedded systems (performance gap).

Another issue that will restrict performance improvements in those systems is the limit in the critical path of the pipeline stages: Intel's Pentium 4 microprocessor has only 12 fanout-of-four (FO4) gate delays per stage, leaving little logic that can be bisected to produce even higher clocked rates. This becomes even worse considering that the delay of those FO4 will increase comparing to other circuitry in the system [1]. One already can see this trend in the newest Intel processors based on the Core and Core2 architectures, which have less pipeline stages than the Pentium 4.

Additionally, one should take into account that the potentially largest problem is excessive power consumption. Still according to [1], future embedded systems must not exceed 75 mW, since batteries do not have an equivalent Moore's law. As previously stated about performance, power spent in future systems is far from the

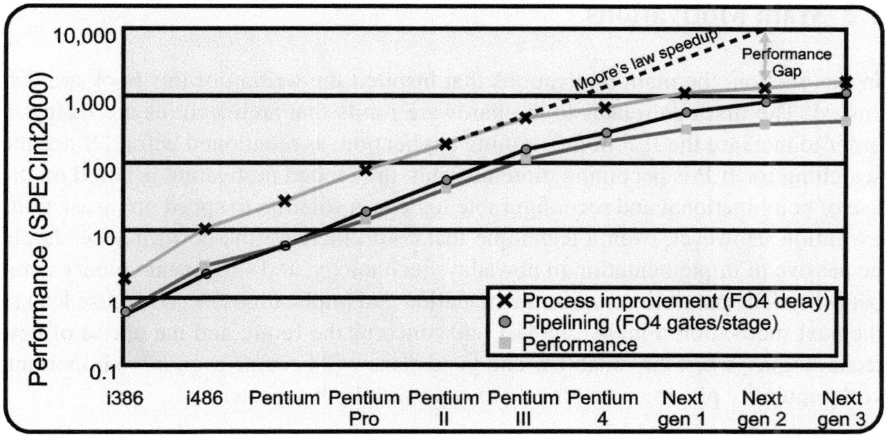

Fig. 1.2 Near future limitations in performance

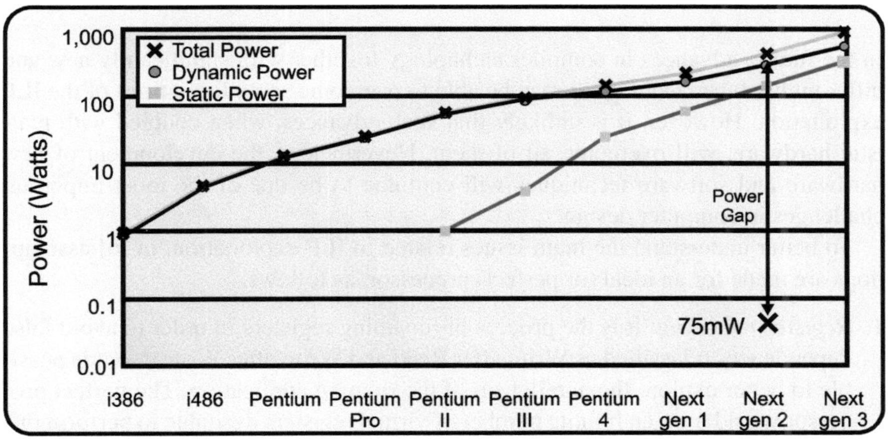

Fig. 1.3 Power consumption in present and future desktop processors

expected, as it can be observed in Fig. 1.3. Furthermore, leakage power is becoming more important and, while a system is in standby mode, it will be the dominant source of power consumption. Nowadays, in general purpose microprocessors, the leakage power dissipation is between 20 and 30 W (considering a total of 100 W) [14].

This way, one can observe that companies are migrating to chip multiprocessors to take advantage of the extra area available, even though, as this book will show, there is still a huge potential to speed up a single thread software. In the essence, the clock frequency increase stagnation, excessive power consumption and higher hardware costs to ILP exploitation, together with the foreseen slower technologies that will be used are new architectural challenges to be dealt with.

1.2 Main Motivations

In this section, the main motivations that inspired the writing of this book are discussed. The first one relates to the hardware limits that architectures are facing in order to increase the ILP of the running application, as mentioned before. Since the searching for ILP is becoming more difficult, the second motivation is based on the use of combinational and reconfigurable logic as a solution to speed up instructions execution. However, even a technique that could increase the performance should be passive of implementation in nowadays technology, and still sustain binary compatibility. The possibilities of implementation and implications of code reuse lead to the next motivation. Finally, the last one concerns the future and the uprise of new technologies, when the reliability and yield costs will become even more important, with regularity playing a major role to cope with both aspects.

1.2.1 Overcoming Some Limits of the Parallelism

In the future, advances in compiler technology together with significantly new and different hardware techniques may be able to overcome some limitations of the ILP exploitation. However, it is unlikely that such advances, when coupled with realistic hardware, will overcome all of them. Nevertheless, the development of new hardware and software techniques will continue to be one of the most important challenges in computer design.

To better understand the main issues related to ILP exploitation, in [6] assumptions are made for an ideal (or perfect) processor, as follows:

1. *Register renaming*: It is the process of renaming registers in order to avoid false dependences (classified as Write after Read and Write after Write), so it is possible to better explore the parallelism of the running application. The perfect processor would have an infinite number of virtual registers available to perform this task and hence all false dependences could be avoided. Therefore, an unbounded number of data independent instructions could begin to be simultaneously executed.
2. *Memory-address alias analysis*: It is the process of comparing memory references encountered in instructions. This is used, for example, to guarantee that a store would not be executed out of order, before a load, both pointing to the same address. Some of these references are calculated at run-time and, as different instructions can access the same address of the memory in a different order, data coherence problems could emerge. In the perfect processor, all memory addresses would be precisely known before the actual execution begins, and a load could be moved before a store, once provided that both addresses are not identical.
3. *Branch prediction*: It is the mechanism responsible for predicting if a given branch will be taken or not, depending on where the execution currently is and

based on previous information (in the case of dynamic types). The main objective is to diminish the number of pipeline stalls due to taken branches. It is also used as a part of the speculation mechanism to execute instructions beyond basic blocks. In an ideal processor, all conditional branches would be correctly predicted, meaning that the predictor would be perfect.

4. *Jump prediction*: In the same manner, all jumps would be perfectly predicted. When combined with perfect branch prediction, the processor would have a perfect speculation mechanism and, consequently, an unbounded buffer of instructions available for execution.

While assumptions 3 and 4 would eliminate all control dependences, assumptions 1 and 2 would eliminate all but the true data dependences. Together, they mean that any instruction belonging to the program's execution could be scheduled on the cycle immediately following the execution of the predecessor on which it depends. It is even possible, under these assumptions, for the last dynamically executed instruction in the program to be scheduled on the very first cycle. Thus, this set of assumptions subsumes both control and address speculation and implements them as if they were perfect.

The analysis of the hardware costs to get as close as possible of this ideal processor is quite complicated. For example, let us consider the instruction window, which represents the set of instructions that are examined for simultaneous execution. In theory, a processor with perfect register renaming should have an instruction window of infinite size, so it could analyze all the dependencies at the same time.

To determine whether n issuing instructions have any register dependences among them, assuming all instructions are register-register and the total number of registers is unbounded, one must compare sources and operands of several instructions. Thus, to detect dependences among the next 2000 instructions requires almost four million comparisons to be done in a single cycle. Even issuing only 50 instructions requires 2,450 comparisons. This cost obviously limits the number of instructions that can be considered for issue at once. To date, the window size has been in the range of 32 to 126, which requires over 2,000 comparisons. The HP PA 8600 reportedly has over 7,000 comparators [6].

Another good example to illustrate how much hardware a superscalar design needs to increase the IPC as much as possible is the Alpha 21264 [9]. It issues up to four instructions per clock and initiates execution on up to six (with significant restrictions on the instruction type, e.g., at most two load/stores), supports a large set of renaming registers (41 integer and 41 floating point, allowing up to 80 instructions in-flight), and uses a large tournament-style branch predictor. Not surprisingly, half of the power consumed by this processor is related to the ILP exploitation [20].

Other possible implementation constraints in a multiple issue processor, besides the aforementioned ones, include: issues per clock, functional units latency and queue size, number of register file ports, functional unit queues, issue limits for branches, limitations on instruction commit, etc.

1.2.2 Taking Advantage of Combinational and Reconfigurable Logic

There are always potential gains when changing the execution mode from sequential to combinational logic. Using a combinational mechanism could be a solution to speed up the execution of sequences of instructions that must be executed in order, due to data dependencies. This concept is better explained with a simple example. Let us have an $n_x n$ bit multiplier, with input and output registers. By implementing it with a cascade of adders, one might have the execution time, in the worst case, as follows:

$$T_{multcombinational} = t_{ppFF} + 2 * n * t_{cell} + t_{setFF} \tag{1.1}$$

where t_{cell} is the delay of an AND gate plus a 2-bits full-adder, $tppFF$ the time propagation of a Flip-Flop, and t_{setFF} the set time of the Flip-Flop.

The area of this multiplier is

$$A_{combinational} = n^2 * A_{cell} + A_{registers} \tag{1.2}$$

considering A_{cell} and $A_{registers}$ as the area occupied by the two bit multiplier cell and registers, respectively.

If one could do the same multiplier by the classical shift and add algorithm, and assuming a carry propagate adder, the multiplication time would be

$$T_{multsequential} = n * (t_{ppFF} + n * t_{cell} + t_{setFF}) \tag{1.3}$$

And the area given by

$$A_{sequential} = n * A_{cell} + A_{control} + A_{registers} \tag{1.4}$$

with $A_{control}$ being the area overhead due to the control unit.

Comparing equations (1.1) with (1.2), and (1.3) with (1.4), it is clear that by using a sequential circuit one trades area by performance. Any circuit implemented as a combinational circuit will be faster than a sequential one, but will most certainly take much more area.

Therefore, the main idea on using reconfigurable hardware is to somehow take advantage of the speedups presented by using combinational logic to perform a given computation. According to [17], with reconfigurable systems, developers can implement circuits that have the potential of being hundreds of times faster than conventional microprocessors. Besides the aforementioned advantage of using a more efficient circuit implementation, the origin of these huge speedups also comes from the circuit's concurrency at various levels (bit, arithmetic and so on). Certain types of applications, which involve intensive computations, such as video and audio processing, encryption, compression, etc are the best candidates for optimization using reconfigurable logic. The programming paradigm is changed, though. Instead of thinking just about temporal programming (one instruction coming after another), it is also necessary to consider spatial oriented models. Considering that reconfigurable systems can be programmed the same way software is to be executed on processors, the author in [16] claims that the hardware is "softening".

This subject will be better explored and explained latter in this book.

1.2.3 Software Compatibility and Reuse of Existent Binary Code

Among thousands of products launched every day, one can observe those which become a great success and those which completely fail. The explanation perhaps is not just about their quality, but it is also about their standardization in the industry and the concern of the final user on how long the product he is acquiring will be subject to updates.

The x86 architecture is one of these major examples. Considering nowadays standards, the X86 ISA (Instruction Set Architecture) itself does not follow the last trends in processor architectures. It was developed at a time when memory was considered very expensive and developers used to compete on who would implement more and different instructions in their architectures. Its ISA is a typical example of a traditional CISC machine. Nowadays, the newest X86 compatible architectures spend extra pipeline stages plus a considerable area in control logic and microprogrammable ROM just to decode these CISC instructions into RISC like ones. This way, it is possible to implement deep pipelining and all other high performance RISC techniques maintaining the x86 instruction set and, consequently, backward compatibility.

Although new instructions have been included in the x86 original instruction set, like the SIMD MMX and SSE ones [4], targeted to multimedia applications, there is still support to the original 80 instructions implemented in the very first X86 processor. This means that any software written for any x86 in any year, even those launched at the end of seventies, can be executed on the last Intel processor. This is one of the keys to the success of this family: the possibility of reusing the existing binary code, without any kind of modification. This was one of the main reasons why this product became the leader in its market. Intel could guarantee to its consumers that their programs would not be surpassed during a long period of time and, even when changing the system to a faster one, they would still be able to reuse and execute the same software again.

Therefore, companies such as Intel and AMD keep implementing more power consuming superscalar techniques and trying to push the frequency increase for their operation to the extreme. More accurate branch predictors, more advanced algorithms for parallelism detection, or the use of Simultaneous Multithreading (SMT) architectures like the Intel Hyperthreading [8], are some of the known strategies. However, the basic principle used for high performance architectures is still the same: superscalarity. While the x86 market is expanding even more, one can observe a decline in the use of more elegant and efficient instruction set architectures, such as the Alpha and the PowerPC processors.

1.2.4 Increasing Yield and Reducing Manufacture Costs

In [11], a discussion is made about the future of the fabrication processes using new technologies. According to it, standard cells, as they are today, will not exist anymore. As the manufacturing interface is changing, regular fabrics will soon become a necessity. How much regularity versus how much configurability (as well as the granularity of these regular circuits) is still an open question. Regularity can be understood as the replication of equal parts, or blocks, to compose a whole. These blocks can be composed of gates, standard-cells, standard-blocks and so on. What is almost a consensus is the fact that the freedom of the designers, represented by the irregularity of the project, will be more expensive in the future. By the use of regular circuits, the design company will decrease costs, as well as the possibility of manufacturing faults, since the reliability of printing the geometries employed today in 65 nanometers and below is a big issue. In [2] it is claimed that maybe the main focus for researches when developing a new system will be reliability, instead of performance.

Nowadays, the resources to create an ASIC design of moderate high volume, complexity and low power, are considered very high. Some design companies can do it because they have experienced designers, infrastructure and expertise. However, for the same reasons, there are companies that just cannot afford it. For these companies, a more regular fabric seems the best way to go as a compromise using an advanced process. As an example, in 1997 there were 11,000 ASIC design startups. This number dropped to 1,400 in 2003 [18]. The mask cost seems to be the primary problem. The estimative in 2003 for the ASIC market is that it had 10,000 designs per year with a mask cost of $20,000. The mask cost for 90-nanometer technology is around $2 million. This way, to maintain the same number of ASIC designs, their costs need to return to tens of thousands of dollars.

The costs concerning the lithography toolchain to fabricate CMOS transistors is one of the major responsible for the high expenses. According to [14], the costs related to lithography steppers increased from $10 to $35 million in this decade, as can be observed in Fig. 1.4. Therefore, the cost of a modern factory varies between $2 and $3 billion. On the other hand, the cost per transistor decreases. Even though it is more expensive to build a circuit nowadays, more transistors are integrated onto one die.

Moreover, it is very likely that the cost of doing the design and verification is growing in the same proportion, increasing even more the final cost. Table 1.1 shows sample non-recurring engineering (NRE) costs for different CMOS IC technologies [18]. At 0.8 mm technology, the NRE costs were only about $40,000. With each advance in IC technology, the NRE costs have dramatically increased. NRE costs for 0.18 mm design are around $350,000, and at 0.13 mm, the costs are over $1 million. This trend is expected to continue at each subsequent technology node, making it more difficult for designers to justify producing an ASIC using nowadays technologies.

Furthermore, the time it takes a design to be manufactured at a fabrication facility and returned to the designers in the form of an initial IC (turnaround time) has also

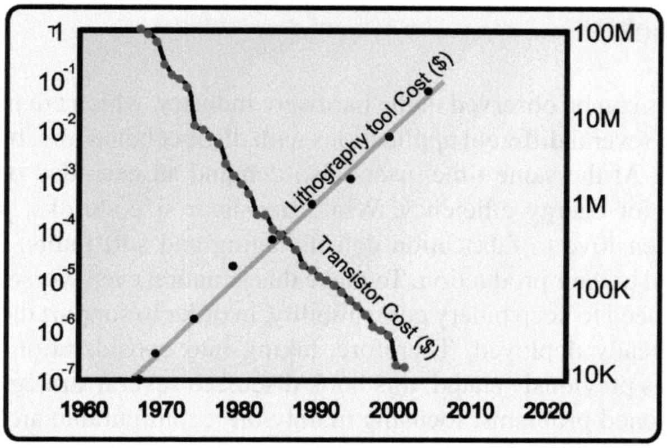

Fig. 1.4 Power consumption in present and future desktop processors

Table 1.1 IC NRE costs and turnaround

	Technology (μm)			
	0.8	0.35	0.18	0.13
NRE (K)	\$40	\$100	\$350	\$1000
Turnaround (days)	42	49	56	76

increased. Table 1.1 provides the turnaround times for four technology nodes. They have almost doubled between 0.8 and 0.13 mm technologies. Longer turnaround times lead to larger design costs, and even possible loss of revenue if the design is late to the market.

Because of all these reasons discussed before, there is a limit in the number of situations that can justify producing designs using the latest IC technology. In 2003, less than 1,000 out of every 10,000 ASIC designs had high enough volumes to justify fabrication at 0.13 mm [18]. Therefore, if design costs and times for producing a high-end IC are becoming increasingly large, just few of them will justify their production in the future. The problems of increasing design costs and long turnaround times are made even more noticeable due to increasing market pressures. The time during which a company seeks to introduce a product into the market is shrinking. This way, the designs of new ICs are increasingly being driven by time-to-market concerns.

Nevertheless, there will be a crossover point where, if the company needs a more customized silicon implementation, it needs be to able to afford the mask and production costs. However, economics are clearly pushing designers toward more regular structures that can be manufactured in larger quantities. Regular fabric would solve the mask cost and many other issues such as printability, extraction, power integrity, testing, and yield.

1.3 This Book

Different trends can be observed in the hardware industry, which are presently being required to run several different applications with distinct behaviors, becoming more heterogeneous. At the same time, users also demand an extended operation, with extra pressure for energy efficiency. While transistor size shrinks, processors are getting more sensitive to fabrication defects, aging and soft faults, increasing the costs associated to their production. To make this situation even worse, designers are stuck with the need to keep binary compatibility, in order to support the huge amount of software already deployed. Therefore, taking into consideration all the issues and motivations previously stated, this book discusses several strategies for solving the aforementioned problems, focusing mainly on reconfigurable architectures and dynamic optimizations techniques.

Chapter 2 discusses the principles related to reconfigurable systems. The potential of executing sequences of instructions in pure combinational logic is also shown. Moreover, a high-level comparison between two different types of reconfigurable systems is performed, together with a detailed analysis of the programs that could be executed on these architectures. Chapter 3 presents a large number of examples of these reconfigurable systems, with a critical analysis of their classification and employed benchmarks. At the end of this chapter it is demonstrated that most of these architectures can present performance boosts just on a very specific subset of benchmarks, which does not reflect the reality of the whole set of applications both embedded and general purpose systems are executing in these days. Therefore, in Chap. 4 two techniques related to dynamic optimization are presented in details: dynamic reuse and binary translation. In Chap. 5, studies that already use both reconfigurable systems and dynamic optimization combined together are discussed. Chapter 6 presents a deeper analysis of one of these techniques, showing a quantitative study on performance, power, energy and area. Finally, the last chapter discusses future work and trends regarding the subjects previously studied, concluding this book.

References

1. Austin, T., Blaauw, D., Mahlke, S., Mudge, T., Chakrabarti, C., Wolf, W.: Mobile supercomputers. Computer **37**(5), 81–83 (2004). doi:10.1109/MC.2004.1297253
2. Burger, D., Goodman, J.R.: Billion-transistor architectures: There and back again. Computer **37**(3), 22–28 (2004). doi:10.1109/MC.2004.1273999
3. Burns, J., Gaudiot, J.L.: Smt layout overhead and scalability. IEEE Trans. Parallel Distrib. Syst. **13**(2), 142–155 (2002). doi:10.1109/71.983942
4. Conte, G., Tommesani, S., Zanichelli, F.: The long and winding road to high-performance image processing with mmx/sse. In: CAMP'00: Proceedings of the Fifth IEEE International Workshop on Computer Architectures for Machine Perception (CAMP'00), p. 302. IEEE Computer Society, Los Alamitos (2000)
5. Flynn, M.J., Hung, P.: Microprocessor design issues: Thoughts on the road ahead. IEEE Micro **25**(3), 16–31 (2005). doi:10.1109/MM.2005.56

6. Hennessy, J.L., Patterson, D.A.: Computer Architecture: A Quantitative Approach, 4th edn. Morgan Kaufmann, San Mateo (2006)
7. Kim, N.S., Austin, T., Blaauw, D., Mudge, T., Flautner, K., Hu, J.S., Irwin, M.J., Kandemir, M., Narayanan, V.: Leakage current: Moore's law meets static power. Computer **36**(12), 68–75 (2003). doi:10.1109/MC.2003.1250885
8. Koufaty, D., Marr, D.T.: Hyperthreading technology in the netburst microarchitecture. IEEE Micro **23**(2), 56–65 (2003)
9. McLellan, E.J., Webb, D.A.: The alpha 21264 microprocessor architecture. In: ICCD'98: Proceedings of the International Conference on Computer Design, p. 90. IEEE Computer Society, Los Alamitos (1998)
10. Prakash, T.K., Peng, L.: Performance characterization of spec cpu2006 benchmarks on Intel core 2 duo processor. ISAST Trans. Comput. Softw. Eng. **2**(1), 36–41 (2008)
11. Rutenbar, R.A., Baron, M., Daniel, T., Jayaraman, R., Or-Bach, Z., Rose, J., Sechen, C.: (when) will fpgas kill asics? (panel session). In: DAC'01: Proceedings of the 38th Annual Design Automation Conference, pp. 321–322. ACM, New York (2001). doi:10.1145/378239.378499
12. Semiconductors, T.I.T.R.: Itrs 2008 edition. Tech. Rep., ITRS (2008). http://www.itrs.net
13. Sima, D.: Decisive aspects in the evolution of microprocessors. Proc. IEEE **92**(12), 1896–1926 (2004)
14. Thompson, S., Parthasarathy, S.: Moore's law: The future of si microelectronics. Mater. Today **9**(6), 20–25 (2006)
15. Thompson, S.E., Chau, R.S., Ghani, T., Mistry, K., Tyagi, S., Bohr, M.T.: In search of "forever," continued transistor scaling one new material at a time. IEEE Trans. Semicond. Manuf. **18**(1), 26–36 (2005). doi:10.1109/TSM.2004.841816
16. Vahid, F.: The softening of hardware. Computer **36**(4), 27–34 (2003). doi:10.1109/MC.2003.1193225
17. Vahid, F.: It's time to stop calling circuits "hardware". Computer **40**(9), 106–108 (2007). doi:10.1109/MC.2007.322
18. Vahid, F., Lysecky, R.L., Zhang, C., Stitt, G.: Highly configurable platforms for embedded computing systems. Microelectron. J. **34**(11), 1025–1029 (2003)
19. Wall, D.W.: Limits of instruction-level parallelism. In: ASPLOS-IV: Proceedings of the Fourth International Conference on Architectural Support for Programming Languages and Operating Systems, pp. 176–188. ACM, New York (1991). doi:10.1145/106972.106991
20. Wilcox, K., Manne, S.: Alpha processors: A history of power issues and a look to the future. In: Proceedings of the Cool-Chips Tutorial. Held in Conjunction with the International Symposium on Microarchitecture. ACM/IEEE, New York (1999)

Chapter 2
Reconfigurable Systems

Abstract As previously discussed, it is possible to take advantage of reconfigurable computing to overcome the main problems that nowadays architectures are facing. Therefore, this chapter aims to explain the basics of reconfigurable systems. It starts with a basic explanation on how these architectures work, their main principles and steps. After that, the principle of merged instruction is introduced, showing how a reconfigurable unit can increase the IPC and affect the number of instructions issued and executed per cycle. The second part of this chapter starts with an overview on the classification of reconfigurable systems, including granularity, instruction types and coupling. Finally, the chapter presents a detailed analysis of the potential gains that reconfigurable computing can present, discussing the main differences, advantages and drawbacks of fine and coarse grain reconfigurable units.

2.1 Introduction

Reconfigurable systems are those architectures that have the capability to adapt themselves to a given application, providing some kind of hardware specialization to it. Through this adaptation, they are expected to achieve great improvements, in terms of performance acceleration and energy savings, when compared to general purpose, fixed instruction set processors. However, because of this certain level of flexibility, the gains are not as high as in Application-Specific Instruction Set Processors (ASIPs) [30] or Application-Specific Integrated Circuits (ASICs) [36].

As an example, let us consider an old ASIC, the STA013. It is an MP3 decoder produced by ST Microelectronics few years ago. It can decode music, at real time, running at 14.7 MHz. Can one imagine the last Intel General Purpose Processor (GPP) decoding an MP3 at real time with that operating frequency? The chip provided by ST is cheaper, faster and consumes less power than any processor that could perform the same task at real time. However, it cannot do anything more than MP3 decoding. For complex systems found nowadays, with a wide range of different applications being executed on it, the Application-Specific approach would

A.C. Schneider Beck Fl., L. Carro, *Dynamic Reconfigurable Architectures and Transparent Optimization Techniques*,
DOI 10.1007/978-90-481-3913-2_2, © Springer Science+Business Media B.V. 2010

Fig. 2.1 Reconfigurable systems: hardware specialization with flexibility

lead to a huge die size, becoming very expensive, since a large number of hardware components would be necessary. On the other hand, a GPP would be able to execute everything, but it is very likely that it would not satisfy either performance or energy constraints of this system.

Reconfigurable architectures were created exactly to fill the gap between specialized hardware and general purpose processing with generic devices. This way, a reconfigurable architecture can be viewed as an intermediate approach between an Application-Specific hardware and a GPP, as Fig. 2.1 illustrates. A reconfigurable system could be configured according to the task at hand, meeting the aforementioned system constraints with a reasonable area occupation, and still being useful for other general-purpose applications. Hence, as Application-Specific components have specialized hardware that accelerate the execution of the applications they were designed for, a system with reconfigurable capabilities would have almost the same benefit without having to commit the hardware into silicon for just one application: computational structures could be adapted after design, in the same way programmable processors can adapt to application changes.

It is important to discuss why reconfigurable architectures can be useful in another point of view. First, let us remember that current architectures used nowadays are based on the Von Neumann model. The problem there is that the Von Neumann model is control-driven, meaning that its execution is based on the program counter. This way, these architectures are still withheld by the so-called Von Neumann bottleneck. Besides representing the data traffic problem, it also has kept people tied to word-at-a-time thinking, instead of encouraging one to think in terms of the larger conceptual units of the task at hand. In contrast, dataflow machines are data-driven: the execution of a given part of the software starts soon after the data required for such operation is ready, so they can explore the maximum parallelism available in the application. However, the employment of dataflow machines implies in the use of special compilers or tools and, most importantly, it changes the programming paradigm. The greatest advantage of reconfigurable architectures is that they can merge both concepts, making possible the use of the very same principle of dataflow architectures, but still using already available tools and compilers, maintaining the programming paradigm.

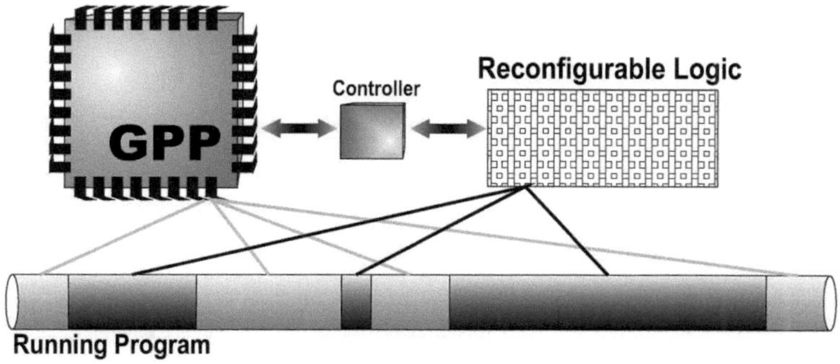

Fig. 2.2 The basic principle of a system making use of reconfigurable logic

2.2 Basic Principles

As already discussed, a reconfigurable architecture is the system that has the ability to adapt itself to perform several and different hardware computations, according to the needs of a given program. This program will not be necessarily always the same. In Fig. 2.2, the basic principle of a computational system working together with a reconfigurable hardware is illustrated. Usually, it is comprised of a reconfigurable logic implemented in hardware, a special component to control and reconfigure it (sometimes it is also responsible for the communication mechanism), a context memory to keep the configurations, and a GPP. Pieces of code are executed on reconfigurable logic (gray), while others are executed by the GPP (dark). The main challenge is to find the best tradeoff considering which pieces of code should be executed on reconfigurable logic. The more software is being executed on reconfigurable logic the better, since it is being executed in a more efficient manner. However, there is a cost associated to it: the need for extra area and memory, which are obviously limited resources.

Systems provided of reconfigurable logic are often called Reconfigurable Instruction Set Processors (RISP) [22], and they will be the focus of this and the next chapters. The reconfigurable logic includes a set of programmable processing units, which can be reconfigured in the field to implement logic operations or functions, and programmable interconnections between them.

2.2.1 Reconfiguration Steps

To execute a program taking advantage of the reconfigurable logic, usually the following steps are necessary (illustrated in Fig. 2.3):

1. *Code Analysis*: the first thing to do is to identify parts of the code that can be transformed for execution on the reconfigurable logic. The goal of this step is to find the best tradeoff considering performance and available resources regarding

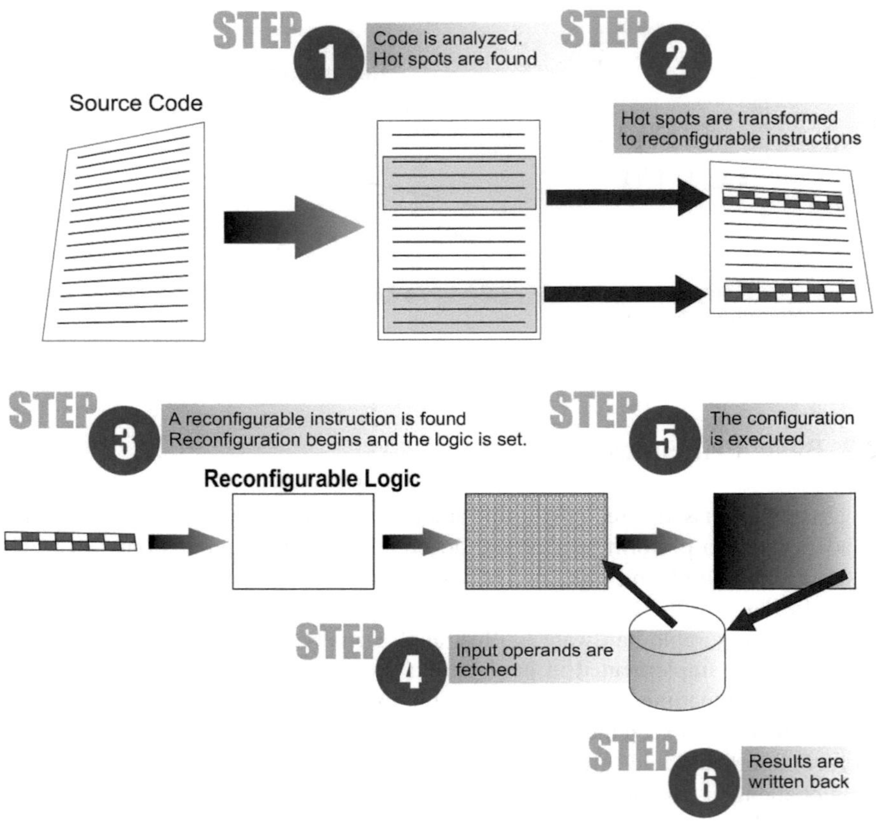

Fig. 2.3 Basic steps in a reconfigurable system

the reconfigurable unit (RU). Usually, the code is not analyzed statically: an execution trace that was previously generated is employed, so dynamic information can be extracted, since it is hard to figure (sometimes impossible) the most executed kernels by just analyzing the source or assembly code. This step can be performed either by automated tools or manually by the designer.

2. *Code transformation*: Once the best candidate parts of code to be accelerated (named as *hot spots* or *kernels*) are found, they need to be replaced by reconfigurable instructions. The reconfigurable instructions will be handled by the control unit of the reconfigurable system. The source code of the processor can also be modified to explicitly communicate with the reconfigurable logic, using native processor instructions.
3. *Reconfiguration*: After code transformation, it is time to send it to the reconfigurable system. When a reconfigurable instruction is found, the programmable components of the reconfigurable logic are organized as a function according to that instruction. This is achieved by downloading from a special memory a set of configuration bits, called *configuration context*. The time needed to configure the

whole system is called reconfiguration time, while the memory required for storing the reconfiguration data is called context memory. Both the reconfiguration time and context memory constitute the reconfiguration overhead.

4. *Input Context Loading*: To perform a given reconfigurable operation, a set of inputs is necessary. They can come from the register file, a shared memory or even be transmitted using message passing.
5. *Execution*: After the reconfigurable unit is set and the proper input operands are ready, execution begins. The operation will be executed in a more efficient manner in comparison with the execution on a GPP.
6. *Write back*: The results of the reconfigurable operation are saved back to the register file, to the memory or transmitted from the reconfigurable unit to the reconfigurable control unit or GPP.

Steps 3 to 6 are repeated while reconfigurable instructions are found in the code, until the end of its execution.

2.3 Underlying Execution Mechanism

To understand how the gains are obtained by the employment of reconfigurable logic, let us start with a very simple example, considering that one wants to build a circuit to multiply a given number by the constant seven. For that, the designer has only two available components: adders and registers. The first choice is to use just one adder and one register (Fig. 2.4a). The result would be generated by repeating seven times the sum operation, so six cycles would be necessary, considering that the register had been reset at the beginning of the operation.

Another choice is to completely replace sequential for combinational logic, eliminating the register and putting six adders directly connected to each other

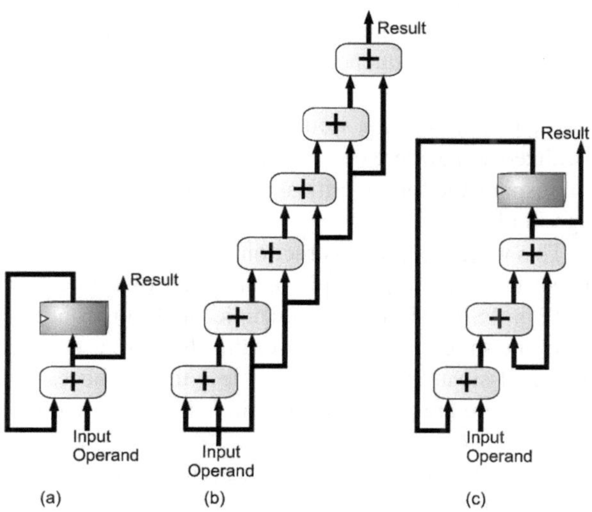

Fig. 2.4 Different ways of performing the same computation

(Fig. 2.4b). The critical path of the circuit will increase, thereby increasing the clock period of the system. However, when considering the total execution time, the second option will be faster, since setup and hold times of the register have been removed. In a certain way, this represents the difference between control and data driven executions commented before. In the first case, the next computation will be performed at the next cycle. In the second case, the next computation will start soon after the previous one was ready.

One could write that the Execution Time (ET) for an algorithm mapped to hardware is

$$A_{sequential} = n * A_{cell} + A_{control} + A_{registers} \tag{2.1}$$

$$ET = number_{cycles} * cycle_{time} \tag{2.2}$$

And for the hardware algorithm of figure (Fig. 2.4a) one has

$$ET_a = 6 * [Tp_{FF} + T_{adder} + T_{set}] \tag{2.3}$$

and for (Fig. 2.4b) one has

$$ET_b = 1 * [6 * T_{adder}] \tag{2.4}$$

and one immediately verifies the second case is faster because the delays of the flip-flops are not in the critical path. However, since one is dealing with combinational logic, one could further optimize by substituting the adder chain by an adder tree, as in Fig. 2.4c, and hence the new execution time would be given by

$$ET_c = 2 * [3 * T_{adder}] \tag{2.5}$$

This would be a compromise of both aforementioned examples. However, the main idea remains the same: to replace, in some level, sequential for combinational logic to group a sequence of operations (or instructions) together. It is interesting to note that in real life circuits, sometimes putting more combinational data to work in a sequential fashion would not increase the critical path, since this path could be localized somewhere else. In some processors, for example, the functional units are not responsible for the critical path of the circuit, so grouping them together may be a good idea.

This way, grouping instructions together to be executed in a more efficient mechanism is the main principle of any kind of application specific hardware, such as ASIP or ASIC. More area is occupied and, consequently, more power is spent. However, one should note that fewer flip-flops are used, and these are a major source of power dissipation. Moreover, as less time is necessary to compute the operations (hence there are performance gains), it is very likely that there will be also energy savings.

Now, let us take the same adder chain presented before, and replace the adders for complete ALUs. Besides, different values can be used as input for these new

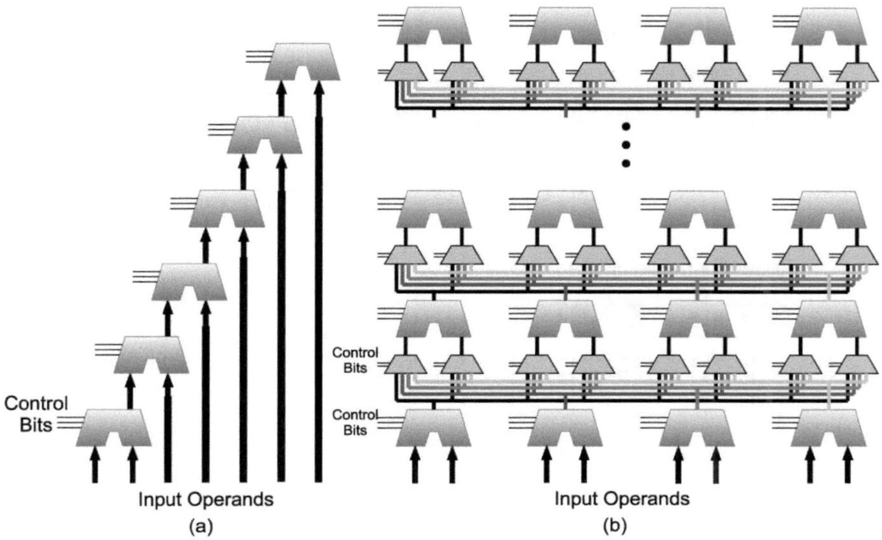

Fig. 2.5 Principles of reconfiguration

ALUs, as it can be observed in Fig. 2.5a. More area is being spent, and it is very likely that the circuit will not be fast as it was before (for example, at least one multiplexer was added at the end of the ALU to select which operation to send as output). Moreover, more control circuitry is necessary (to configure the ALUs). On the other hand, now there is certain flexibility: any arithmetic and logic operation can be performed. Extending this concept even more, it is possible to add ALUs working in parallel, and multiplexers to route the values between them (Fig. 2.5b). Again, the critical path increases, even more control hardware is necessary, but there is still more flexibility, besides the possibility of executing operations in parallel. The main principle remains the same: to group instructions to be executed in a more efficient manner, but now with some flexibility. This is, in fact, an example of a coarse grain reconfigurable array, and it will be seen in more details later in this chapter.

Figure 2.6 graphically shows the difference between using reconfigurable logic and a traditional parallel architecture to execute instructions. The upper part of the figure demonstrates the execution of several instructions on a traditional parallel architecture, such as the superscalar ones. These instructions are represented as boxes. Those that have the same texture represent instructions that are data dependent and hence cannot be executed in parallel, while non-dependent instructions can be executed concurrently. There is a limit, though: no matter how many functional units are available, sequences of dependent instructions must be executed in order. On the other hand, by using the data-driven approach and combinational logic, one is able to reduce the time spent by executing exactly the sequences of dependent instructions in a more efficient manner (avoiding the flip-flop delays in the reconfigurable logic),

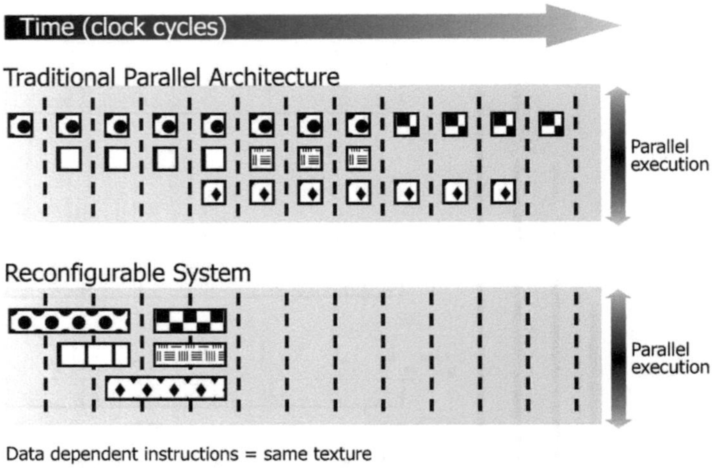

Fig. 2.6 Performance gains obtained when using reconfigurable logic

at the cost of extra area. Consequently, as a legacy of dataflow machines, reconfigurable systems, besides being able to explore the parallelism between instructions, can also speed up instructions which are data dependent between themselves, in opposite to traditional architectures.

2.4 Advantages of Using Reconfigurable Logic

The widely used Patterson [27] metrics of relative performance through measures such as IPC (Instructions Per Cycle) are well suited for comparing different processor technologies and ISA (Instruction Set Architecture), as it abstracts concepts such as clock frequency. As described in [34], however, to better understand the performance evolution in the microprocessor industry, it is interesting to note the *Absolute Processor Performance* (*Ppa*) metric denoted as:

$$Ppa = fc * 1/CPII * IPII * OPI(operations/sec) \tag{2.6}$$

In (2.6), *CPII*, *IPII* and *OPI* are described respectively as *Cycles Per Issue Interval*, *Instructions Per Issue Interval* and *Operation per Instructions*, while *fc* is the operating clock frequency. The first two metrics, when multiplied, form the known *IPC* rate. Nevertheless, it is interesting to keep these factors separated in order to better expose speed-up potentials.

The *CPII* rate informs the intrinsic temporal parallelism of the microarchitecture, showing how frequently new instructions are issued to execution. The *IPII* variable is related to the issue parallelism, or the average number of dynamically fetched instructions issued to execution per issue interval. Therefore, temporal (*CPII*) and issue (*IPII*) parallelisms can be illustrated by the following equations:

$$IPII = (Number\ of\ Instructions)/(Number\ of\ issues) \tag{2.7}$$

$$CPII = (Number\ of\ Cycles)/(Number\ of\ Issues) \tag{2.8}$$

Finally, the *OPI* metric measures intra-instruction parallelism, or the number of operations that can be issued through a single binary instruction word. It is important to notice that one should distinguish the *OPI* from the *IPII* rate, since the first reflects changes in the binary code that should be adapted statically to boost intrainstruction parallelism, such as data parallelism found in SIMD architectures, while the second is related to the number of instructions that are dynamically issued to be executed in parallel, such as the ones sent for execution in a superscalar processor after scheduling. Figure 2.7 illustrates these three metrics.

Throughout the microprocessor evolution history, several approaches have been considered to improve performance by manipulating one or more of the factors of (2.6). One of these approaches, for example, deals with the *CPII* metric by increasing instructions throughput with pipelining [27]. Moreover, the *CPII* metric has also been well covered with efficient branch prediction mechanisms and memory hierarchies, though this metric is still limited by pipeline stalls such as the ones caused by cache misses. The *OPI* rate has been dealt with the development of complex CISC instructions or SIMD architectures. On the other hand, few solutions other than the superscalar approach since the 90's explored the opportunity of increasing the *IPII* rate.

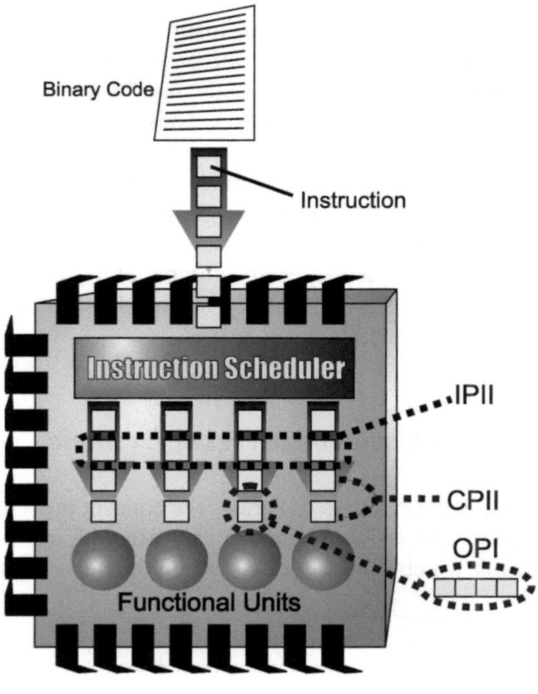

Fig. 2.7 Gains obtained when using reconfigurable logic

2.4.1 Application

A reconfigurable system targets to increase exactly the *IPII* rate. As can be observed in (2.7), in order to increase the *IPII* number, it is necessary to increase the execution efficiency by decreasing the number of issues. Considering that a sequence of instructions is identified and grouped to be executed on the reconfigurable system, more instructions will be issued by issue interval (so increasing the *IPII* rate). Equation (2.9) shows how the number of issues is affected by the technique:

$$\mathit{Number\ of\ Issues} = \mathit{Total\ number\ of\ executed\ Instructions} + \mathit{NMI} * (1 - \mathit{AMIL}) \tag{2.9}$$

where the *Average Merged Instructions Length* (*AMIL*) is the average group size in number of instructions; while the *Number of Merged Instructions* (*NMI*) counts how many merged instructions[1] were issued for execution on the combinational logic. This can be represented by the following equation:

$$\mathit{NMI} = \mathit{MIR} * \mathit{Total\ number\ of\ executed\ Instructions} \tag{2.10}$$

MIR is denoted as the *Merged Instructions Rate*. This is an important factor as it exposes the density of grouped operations that can be found in an application. If *MIR* is equal to one, then the whole application was mapped into an efficient mechanism, and there is no need for a processor, which is actually the case of specialized hardware (ASIPs or ASICs) or complete dataflow architectures.

Furthermore, doing a deeper analysis, one can conclude that the ideal *CPII* also equals to one, which means that the functional units are constantly fed by instructions every cycle. However, due to pipeline stalls or to instructions with high delays, the *CPII* variable tends to be of a greater value. In fact, manipulating this factor is a bit more complicated, as both the number of cycles and the number of issues are affected by the execution of instructions on reconfigurable logic. As it will be shown in the example, there are times when the *CPII* will increase; this is actually a consequence of the augmented number of operations issued in a group of instructions.

This way, one thing that must be assured is that the *CPII* rate will not grow in a manner to cancel the *IPC* gains caused by the increase of *IPII*. In other words, if the number of issues decreases, the number of cycles taken to execute instructions also has to decrease. Consequently, a fast mechanism is necessary for reconfiguring the hardware and executing instructions.

2.4.2 An Instruction Merging Example

The following example illustrates the concept previously proposed.

Figure 2.8a shows a hypothetical trace with instructions *a*, *b*, *c*, *d* and *e*, and the cycles at which the instruction execution ends. If one considers that a given GPP

[1] In this chapter the set of instructions that are executed on reconfigurable hardware is called merged instructions; in previous works, several and different names have been used.

architecture has an *IPII* rate of one, typical of RISC scalar architectures, and that *inst d* causes a pipeline stall of 5 cycles (for instance, this instruction must wait for the result of another one, in a typical case of true data dependence), while all other instructions are executed in one cycle, this trace of 14 instructions would take 18 cycles to execute. This results in a *CPII* of 1.28 and an *IPC* of 0.78.

If, however, instructions of number one to five are merged (which is represented by *Inst M*, as shown in Fig. 2.8b, and executed in two cycles, the whole sequence would then be executed in 14 cycles. Note that the left column in Fig. 2.8b represents

Number	Instruction	Cycle
1	inst a	1
2	inst b	2
3	inst b	3
4	inst a	4
5	inst c	5
6	inst b	6
7	inst a	7
8	inst a	8
9	inst b	9
10	inst d	10
11	inst e	15
12	inst b	16
13	inst c	17
14	inst a	18

CPII = 1.28
IPII = 1

IPC* = *0.78

(a)

Issue	Instruction	Cycle
1	***inst M***	1
2	inst b	3
3	inst a	4
4	inst a	5
5	inst b	6
6	inst d	11
7	inst e	12
8	inst b	13
9	inst c	14
10	inst a	15

CPII = 1.5
IPII = 1.4

IPC* = *1

(b)

Issue	Instruction	Cycle
1	***inst M***	1
2	inst b	3
3	inst a	4
4	inst a	5
5	***inst M2***	8
6	inst b	9
7	inst c	10
8	inst a	11

CPII = 1.375
IPII = 1.75

IPC* = *1.27

(c)

Fig. 2.8 (**a**) Execution trace of a given application; (**b**) Trace with one merged instruction; (**c**) Trace with two merged instructions

the issue number of the instruction group. Therefore, one would find the following numbers: $CPII = 1.5, AMIL = 5$, and $MIR = 1/14 = 0.07$. Because of the capability of speeding up the fetch and execution of the merged instructions, the final *IPII* would increase to 1.4. Even though the *CPII* would increase from 1.28 to 1.5, the *IPC* rate would grow from 0.78 to 1.

Nevertheless, one could expect further improvements if merged instructions included *Inst d*, which caused a stall of 5 cycles in the processor pipeline. Supposing that the sequence of instructions *b*, *d* and *e* (issue numbers of 5, 6 and 7 in Fig. 2.8b) is merged into instruction M2 and executed in 3 cycles, it would produce an impact on the *CPII* that would go down to 1.375 while the *IPII* would rise to 1.75, resulting in an *IPC* equals to 1.27. In this example, the fact of executing these instructions in a dataflow manner would mask the delay effects of data dependency. This is illustrated in Fig. 2.8c. This way, when using a reconfigurable system, the interval of execution between a set of instruction and another can be longer than the usual. However, as more instructions are executed per time slice, *IPC* increases.

Later in this chapter, an ideal solution is analyzed, which is capable of executing merged instructions in just one cycle, meaning that the *CPII* inside the reconfigurable fabric is 1. This will show the potential gains of using reconfigurable logic when affecting the *AMIL* and *IPII* rates.

2.5 Reconfigurable Logic Classification

In the reconfigurable field, there is a great variety of classifications, as it can be observed in some surveys published about the subject [22, 25, 35, 37]. In this book, the most common ones are discussed.

2.5.1 Code Analysis and Transformation

This subject concerns how the best hot spots are found in order to replace them with reconfigurable instructions (transforming the code) and the level of automation of this process.

Code analysis can be done in the binary/source code, or yet in the trace generated from the execution of the program on the target GPP. The designer can find the hot spots analyzing the source code (looking for loops with great number of interactions, for instance), or the trace. The greatest advantage of using the trace is that it contains dynamic information. For instance, the designer cannot know if loops with non-fixed bounds are the most used ones by only analyzing the source code. The designer can also benefit from automated tools to do this job. These tools usually work on the trace and can indicate to the designer which are the most executed kernels.

After the hot spots were found, it is time to replace them with reconfigurable instructions. These instructions are related to the communication, reconfiguration

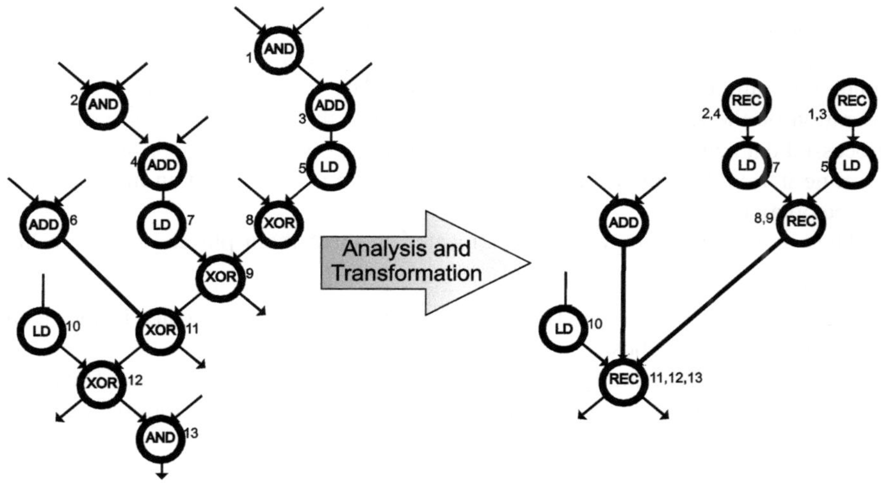

Fig. 2.9 Analysis and transformation of a code sequence based on DFG analysis

and execution processes. Again, the level of automation is variable. It could be the designer's responsibility the whole work of replacing the hotspots with reconfigurable instructions directly in the assembly code. Yet, code annotation can be used. For instance, macros can be employed in the source code to indicate that there will be a reconfigurable instruction. The assembler then will be used to automatically generate the modified code. Finally, there is the complete automated process: given a set of constraints related to a given reconfigurable architecture, a tool will obtain information about the most used hot spots and transform them to reconfigurable instructions, handling issues such as communication between the GPP and reconfigurable logic, reconfiguration overheads, execution and write back of results. It is important to note that such tools are highly dependent to the reconfigurable system they were built to be used with.

Automated tools usually involve some complex graph analysis in order to find the best alternatives for code transformation. To better illustrate this, let us consider an example based on [24], demonstrated in Fig. 2.9. As it can be observed, the sequence of instructions is organized in a DFG (Data Flow Graph). Some sequences are merged together and transformed to a reconfigurable instruction.

These automated tools sometimes can also include another level of code transformations. These happen before code analysis, and are employed to better expose code parallelism, using compiler techniques such as superblock [29] or hyperblock [31].

2.5.2 RU Coupling

How the reconfigurable logic is coupled, or connected to the main processor, defines how the interface between both machines works, including issues related to how data is transferred and how the synchronization between the parts is performed.

The position of the reconfigurable logic, relative to the microprocessor, directly affects performance. The benefit obtained from executing a piece of code on it depends on communication and execution costs. The time necessary to execute an operation on the reconfigurable logic is the sum of the time needed to transfer the processed data and the time required to process it. If this total time is smaller than the time it would normally take in the standalone processor, then an improvement can be obtained.

The reconfigurable logic can be allocated in three main places relative to the processor:

- *Attached to the processor*: The reconfigurable logic communicates to the main processor through a bus.
- *Coprocessor*: The reconfigurable logic is located next to the processor. The communication is usually done using a protocol similar to those used for floating point coprocessors.
- *Functional Unit*: The logic is placed inside the processor. It works as an ordinary functional unit, having full access to the processor's registers. Some part of the processor (usually the decoder) is responsible to activate the reconfigurable logic, when necessary.

Figure 2.10 illustrates these three different types of coupling. The two first interconnection schemes are usually called loosely coupled. The functional unit approach, in turn, is named tightly coupled. As stated before, the efficiency of each technique depends on two things: the time required to transfer data between the components, where, in this case, the functional unit approach is the fastest one and the attached processor, the slowest; and the quantity of instructions executed by the reconfigurable logic. Usually, loosely coupled units can execute larger chunks of code, and are faster than the tightly coupled ones, mainly because they have more area available. For loosely coupled units, there is a need for faster execution times: it is necessary to overcome some of the overhead brought by the high delays caused by the data transfers. The data exchange is usually performed using shared memory, while the communication can be done using shared memory or message passing.

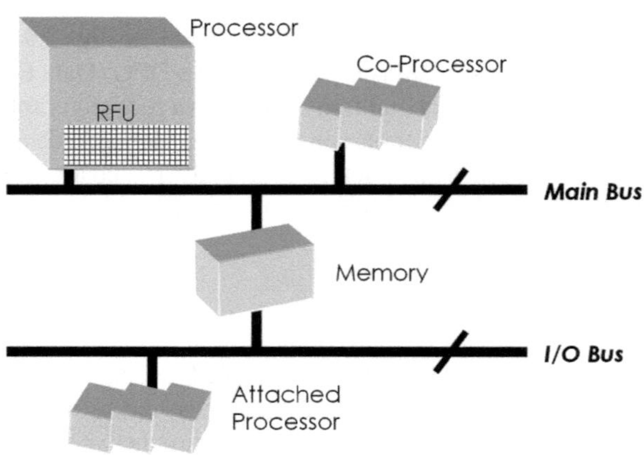

Fig. 2.10 Different types of RU coupling

A tightly coupled RU, although increasing the area taken by the processor itself, makes the control logic simpler. Besides, it minimizes the overhead required in the communication between the reconfigurable array and the rest of the system, because it can share some resources with the processor, such as the access to the register bank, which is usually employed for the communication between the main processor and the reconfigurable unit. On the other hand, the tightly coupled functional unit must run fast enough in order to avoid increasing the processor's cycle time, and hence the amount of logic that can be packed is limited.

When there is a reconfigurable unit working as functional unit in the main processor, it is called a Reconfigurable Functional Unit, or RFU. The first reconfigurable systems were implemented as co-processors, or as attached processors. However, with the manufacturing advances and more transistors available within the same die, the RFU based system is becoming a very common implementation.

2.5.3 Granularity

The granularity of a reconfigurable unit defines its level of data manipulation: the smallest possible parts for reconfiguration (or building blocks) for fine-grained logic are usually gates (efficient for bit level operations), while in coarse-grained RUs these blocks are larger (like ALUs), therefore better suited for bit parallel operations.

A fine-grain reconfigurable system consists of Processing Elements (PEs) and interconnections that are configured at bit-level. The PEs implement any 1-bit logic function and vast interconnection resources are responsible for the communication links between these PEs. Fine-grain systems provide high flexibility and can be used to implement theoretically any digital circuit. They are usually implemented with or based on FPGA. An example of an FPGA architecture is shown in Fig. 2.11. It consists of a 2-D array of Configurable Logic Blocks (CLBs) used to implement both combinational and sequential logics. Each CLB typically consists of a 4-input lookup table (LUT) and a flip-flop. The lookup table is responsible for executing a given logic operation, so it can implement any 1-bit logic function. Programmable interconnects surround CLBs, ensuring the communication between them. These interconnections can be either direct connections via programmable switches (e.g., pass transistors) or a mesh structure using Switch Boxes (S-Box), as illustrated in the example. Finally, programmable I/O cells surround the array, for communication with the external environment.

A coarse-grain reconfigurable system, in turn, consists of reconfigurable PEs that implement word-level operations and special-purpose interconnections retaining enough flexibility to map different applications onto the system. While bit-oriented algorithms can take better benefit from fine-grained systems, the coarse-grain approach may be the best alternative for computation intensive applications. Coarse grain architectures are implemented using off the shelf functional units, or yet special functional units targeted to a given domain of application. The interconnection resources are usually multiplexers, crossbars or buses. Figure 2.12 illustrates a simple reconfigurable array of functional units, connected to each other using crossbars. The word length of this array is 16 bits long.

Fig. 2.11 A typical FPGA architecture

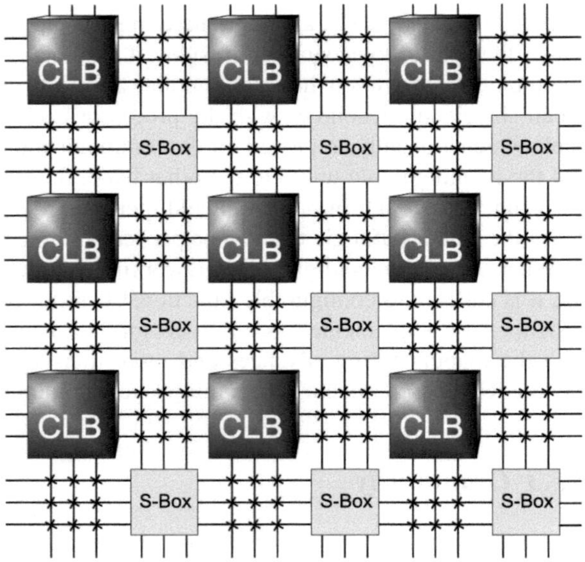

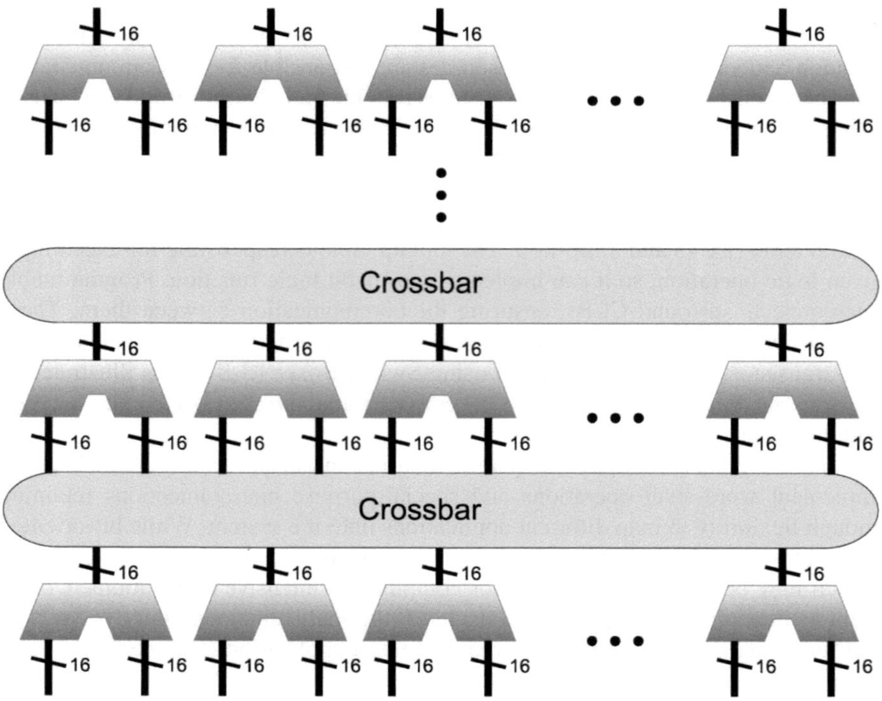

Fig. 2.12 An example of a coarse grain reconfigurable array of functional units

Granularity also affects the size of the configuration context and the configuration time. With fine-grained logic, more information is needed to describe the reconfigurable instruction. Coarse-grained logic descriptions are more compact, but on the other hand, some operations can be limited due to its higher level of data manipulation.

Another issue related to the granularity is the segment size. A segment is the minimum hardware unit that can be configured and assigned to a reconfigurable instruction (which will be explained in the following sub-section). Segments allow reconfigurable instructions to share the reconfigurable resources. If segments are used, the configuration of the reconfigurable logic can be performed in a hierarchical manner. Each instruction is assigned to one or more segments, and inside those segments, the processing elements are configured. The interconnect that connects the elements inside a segment is referred to as intra-segment interconnect. Intersegment interconnect is used to connect different segments.

2.5.4 Instruction Types

Reconfigurable instructions are those responsible for controlling the reconfigurable hardware, as well as for the data transfer between it and the main processor. They are usually identified by special opcodes in the processor instruction set. Which operation a reconfigurable instruction will perform is usually specified using an extra field in the instruction word. This field can give two different kinds of information:

- *Address*: The special field indicates the memory address of the configuration data.
- *Instruction number*: An instruction identifier of small length is embedded in the instruction word. This identifier indexes a configuration table where some information, such as the configuration data address, is stored. The number of reconfigurable instructions is limited to the size of the table.

The first approach needs more instruction word bits, but has the benefit that the number of different instructions is not limited to the size of a table, as in the second case. On the other hand, when using the configuration table approach, the table can be changed on the fly, so the processor can more easily adapt to the task at hand at runtime. However, specialized scheduling techniques have to be used during code generation in order to configure what instructions will be available in the table at a given moment, during program execution.

There are other issues concerning instructions in reconfigurable systems. For example, the memory accesses performed by these instructions can be made by specialized load/store operations or implemented as stream based operations. If the memory hierarchy supports several accesses at the same time, then the number of memory ports can be greater than one. Moreover, the register file accessed by the reconfigurable unit can be shared with other functional units or be dedicated (such as the floating point register file in some architectures). The dedicated register file would need less ports than if it was shared, becoming cheaper to be implemented.

The major drawback of a dedicated register file is that more control for synchronizations is necessary.

In the same way, reconfigurable instructions can be implemented as stream based ones or customized. The first type can process large amounts of data in a sequential or blocked manner. Only a particular set of applications can benefit from this type, such as FIR filtering, discrete cosine transformation (DCT) or other signal processing related algorithms. Custom instructions take small amounts of data at a time (usually from internal registers) and produce another small amount of data. These instructions can be used in almost all applications as they impose fewer restrictions on the characteristics of the application. Example of these operations are bit reversal, multiply accumulate (MAC) etc. Instructions can also be classified in many other ways, such as execution time, pipelining, internal state etc. For more details on these classifications, refer to [25].

2.5.5 Reconfigurability

The logic can be programmed at different moments. If it can only be programmed at startup, before execution begins, this unit is not reconfigurable (it is configurable). If the logic can be programmed after initialization, then it is called reconfigurable. The application can be divided in different blocks of functionality, so the RU can be reconfigured according to the needs of each individual block. In this manner, the program adaptation is done in a per block basis. The reconfigurable logic is simpler to be implemented if the fabric is blocked during reconfiguration. However, if the RU can be used while being reconfigured, it is possible to increase performance. This can be done, for example, by dividing the RU in segments that can be configured independently from each other. The process of reconfiguring just parts of the logic is called partial reconfiguration [25].

Reconfiguration times depend on the size of the configuration data. The configuration data is usually composed of the configuration bits for the unit reconfiguration, as well as information about the input context. These times depend on the configuration method used. For instance, in the PRISC processor [21], the RU is configured by copying the configuration data directly into the configuration memory using normal load/store operations. If this task is performed by a configuration unit that is able to fetch the configuration data while the processor is executing code, a performance gain can be obtained. Hence, prefetching the configuration data can reduce the time the processor is stalled waiting for reconfiguration. The employment of this approach can be done automatically by software tools [32].

2.6 Directions

In this sub-section, some tradeoffs that should be taken into account while developing a reconfigurable architecture are analyzed. First, a known benchmark set is evaluated in order to figure what is the best strategy to take in terms of granularity. Then, the impact of this analysis in both fine and coarse grain reconfigurable systems performing high levels simulations is studied. Finally, other issues are considered, such as reconfiguration and execution times, and the growing number of applications being executed at the same time on a system.

2.6.1 Heterogeneous Behavior of the Applications

In [23] a subset of the Mibench Benchmark Suite [26] is used, which represents a complete set of diverse algorithm behaviors. As a matter of fact, this suite has been chosen because, according to [26], it has a larger range of different behaviors when compared against other benchmark sets, e.g. SPEC2000 [28]. This way, the following 18 benchmarks were evaluated: *Quicksort*, *Susan Corners/Edges/Smoothing*, *Jpeg Encoder/Decoder*, *Dijkstra*, *Patricia*, *StringSearch*, *Rinjdael Encode/Decode*, *Sha*, *Raw Audio Coder/Decoder*, *GSM Coder/Decoder*, *Bitcount* and *CRC32*.

First, a characterization of the algorithms regarding the number of instructions executed per branch is done (classifying them as control or dataflow oriented based on these numbers). As it can be observed in Fig. 2.13, the *RawAudio Decoder* algorithm is the most control flow oriented one (a high rate of executed branches) while the *Rijndael Encoder* is quite the opposite, being data flow oriented. It is important to point out that, for reconfigurable architectures, the more instructions a basic block has, the better, since there is more room for exploiting parallelism. Furthermore, more branches mean additional paths that can be taken, increasing the execution time and the area consumed by a given configuration, when implemented in reconfigurable logic.

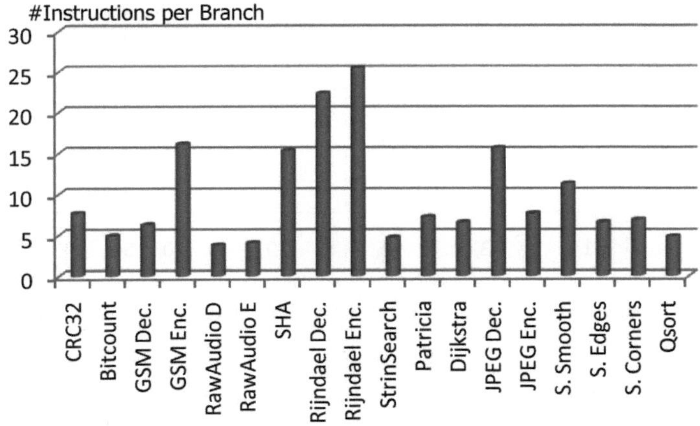

Fig. 2.13 Instruction per Branch Rate

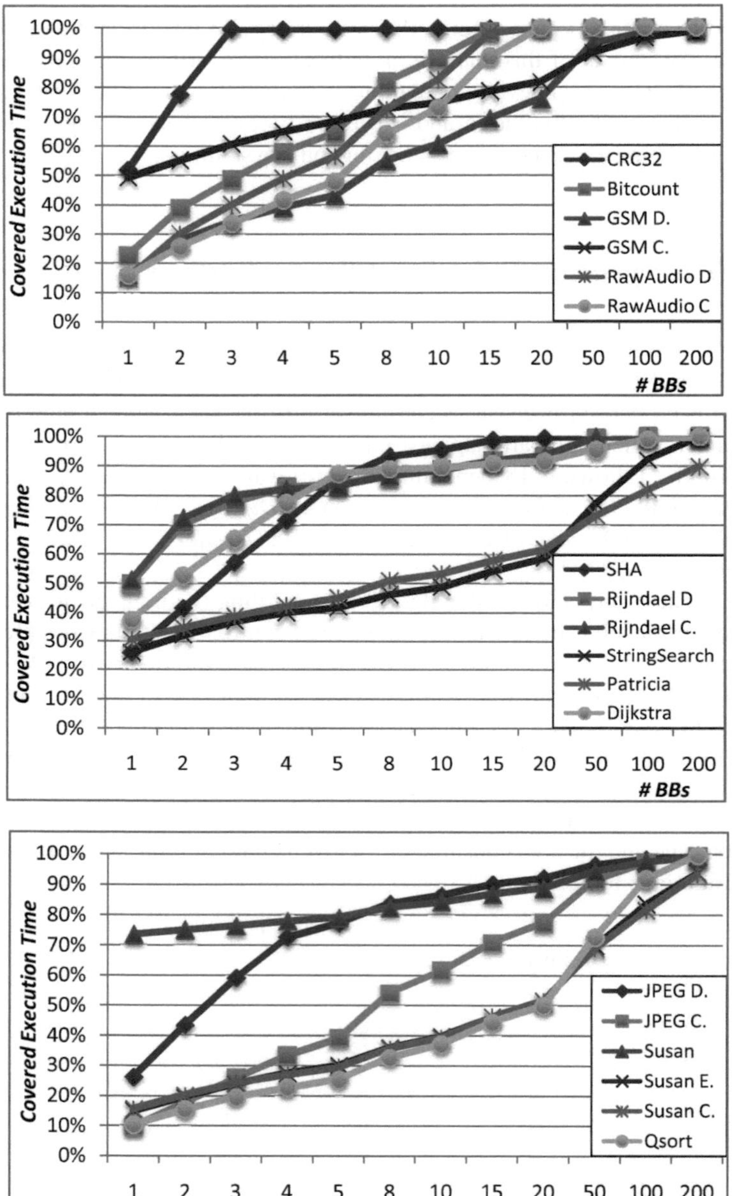

Fig. 2.14 How many BBs are necessary to cover a certain amount of execution time?

Figure 2.14 shows the analysis of distinct kernels based on the execution rates of the basic blocks in the programs. The methodology involves investigating the number of basic blocks responsible for covering a certain percentage of the total number of basic block executed. For instance, in the *CRC32* algorithm, just 3 basic blocks are responsible for almost 100% of the total program execution time. Again, for typical reconfigurable systems, this algorithm can be easily optimized: one just needs to concentrate all the design effort on that specific group of basic blocks and implement them to reconfigurable logic.

However, other algorithms, such as the widely used *JPEG decoder*, have no distinct execution kernels at all. In this algorithm, 50% of the total instructions executed are due to 20 different BBs. Hence, if one wished to have a speedup factor of 2x, according to Amdahl's law, all 20 different basic blocks should be mapped into reconfigurable logic, and each should be accelerated by a factor of 4. This analysis will be presented in more details in the next section.

The problem of not having a clear group of most executed kernels becomes even more evident if one considers the wide range of applications that embedded systems are implementing nowadays. In a scenario when an embedded system runs *RawAudio decoder*, *JPEG encoder/decoder*, and *StringSearch*, the designer would have to transform approximately 45 different basic blocks into the reconfigurable fabric to achieve a maximum of 2 times performance improvement.

Furthermore, it is interesting to point out that the algorithms with a high number of instructions per branch tend to be the ones that need fewer kernels to achieve higher speedups. Figure 2.15 illustrates this scenario by using the cases with 1, 3 and 5 basic blocks. Note that, mainly when one considers only the most executed basic blocks (first bar of each benchmark), the shape of the graph is very similar to the instructions per branch ratios shown in Fig. 2.13 (with some exceptions, such as the *CRC32* or *JPEG decoder* algorithms). A deeper study about this issue could be envisioned to indicate some directions regarding the reconfigurable arrays optimization just based on very simple profile statistics.

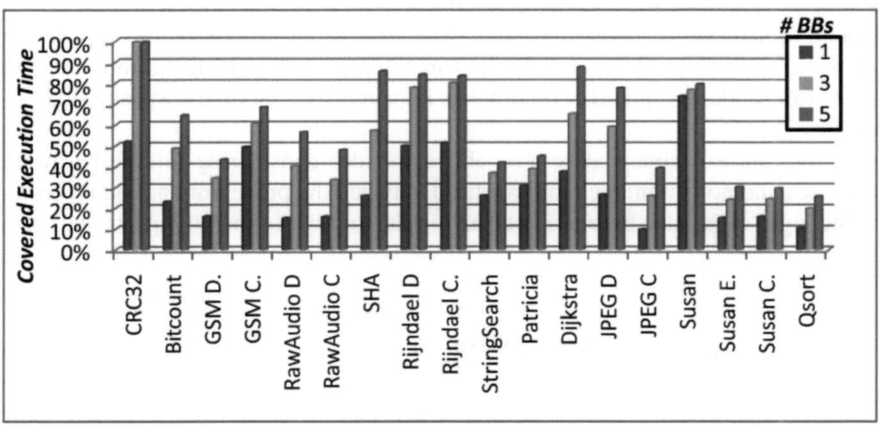

Fig. 2.15 Amount of execution time covered by 1, 3 or 5 basic blocks in each application

2.6.2 Potential for Using Fine Grained Reconfigurable Arrays

In this section, the potentiality of fine grain reconfigurable arrays is evaluated. Considering hot spots as being loops or subroutines, the level of performance gains one can obtain whenever a determined number of them is mapped to a fine grain reconfigurable logic is analyzed. In this first experiment, it is assumed that just one piece of reconfigurable hardware is available per loop or subroutine. This means that the only part of the code that will be optimized by the reconfigurable logic is the one that is common to all iterations. For example, let us assume that a loop is executed 50 times. 100% of the code is executed 49 times, but only 20% is executed 50 times. This way, just this part will be optimized, since it comprises the common part of code executed in all loop iterations. Figure 2.16 illustrates this case. Moreover, subroutines called inside loops are not suited for optimization.

Figures 2.17a and b show, in the y-axis, the performance improvements (speedup factor) when implementing a different amount of subroutines or loops (x-axis) on reconfigurable logic, respectively. The hot spots are chosen in order of relevance, where the first on the list is the most executed one (considering how many times it is repeated and its number of instructions). It is assumed that the execution time for each one of these hot spots would be of one cycle, when reconfigurable hardware is used. Although extremely optimistic, this can give us an upper bound on the execution time. As it can be observed, the performance gains demonstrated are very heterogeneous. For a group of algorithms, just a small number of subroutines or loops implemented on fine grain reconfigurable logic are necessary to show good speedups. For others, the level of optimization is very low.

One of the reasons for the lack of optimization in some algorithms is the methodology used for code allocation on the reconfigurable logic, explained above. This way, even if there is a huge number of hot spots subject to optimization, but presenting different dynamic behaviors, just a small number of instructions inside these hot spots could be optimized. This shows that automatic tools, aimed at searching the best parts of the software to be transformed to reconfigurable logic, might not be enough to achieve the necessary gains, whenever heterogeneity on the application set comes into play. Consequently, human interaction for changing and adapting parts of the code is required, with obvious impact on design time costs and time-to-market.

In the first experiment described above, besides considering infinite hardware resources and no communication overhead between the processor and the reconfigurable logic, it has also been assumed an infinite number of memory ports with zero

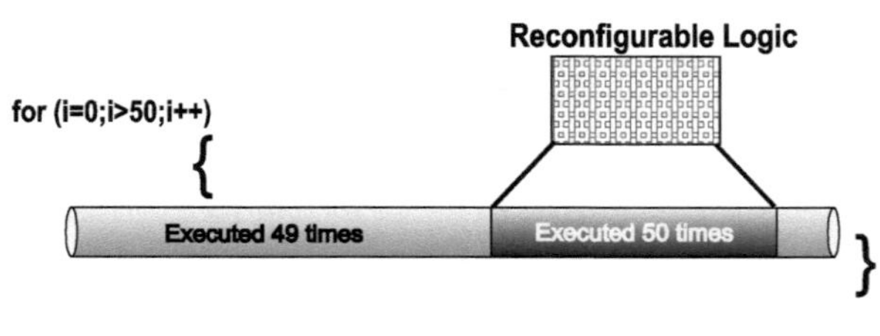

Fig. 2.16 Just a small part of the loop can be optimized

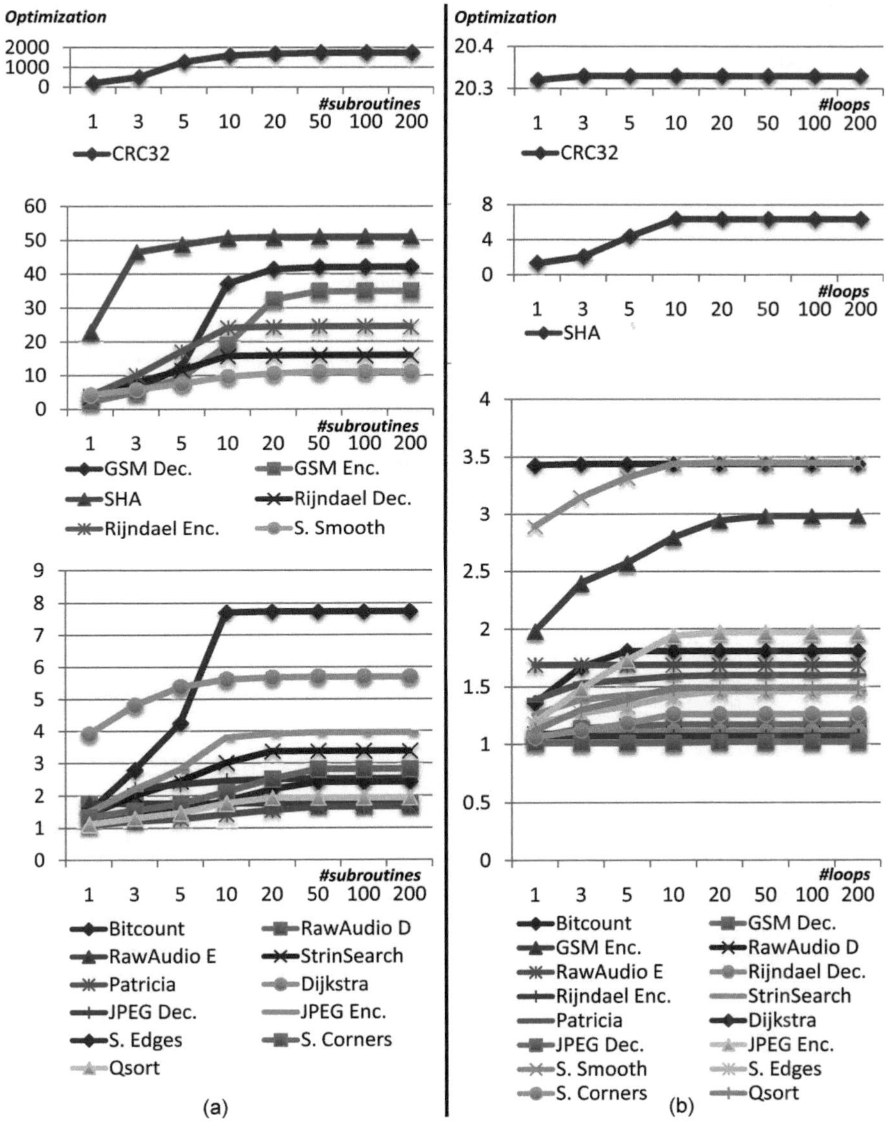

Fig. 2.17 Performance gains considering different numbers of (**a**) subroutines and (**b**) loops being executed in 1 cycle in reconfigurable logic

delay, which is practically infeasible for any relatively complex configuration. Now, in Fig. 2.18, a more realistic assumption is considered: each hot spot would take 5 cycles to be executed on the reconfigurable logic. These extra cycles were added to give a hint on the impact of limited memory ports. When comparing this experiment

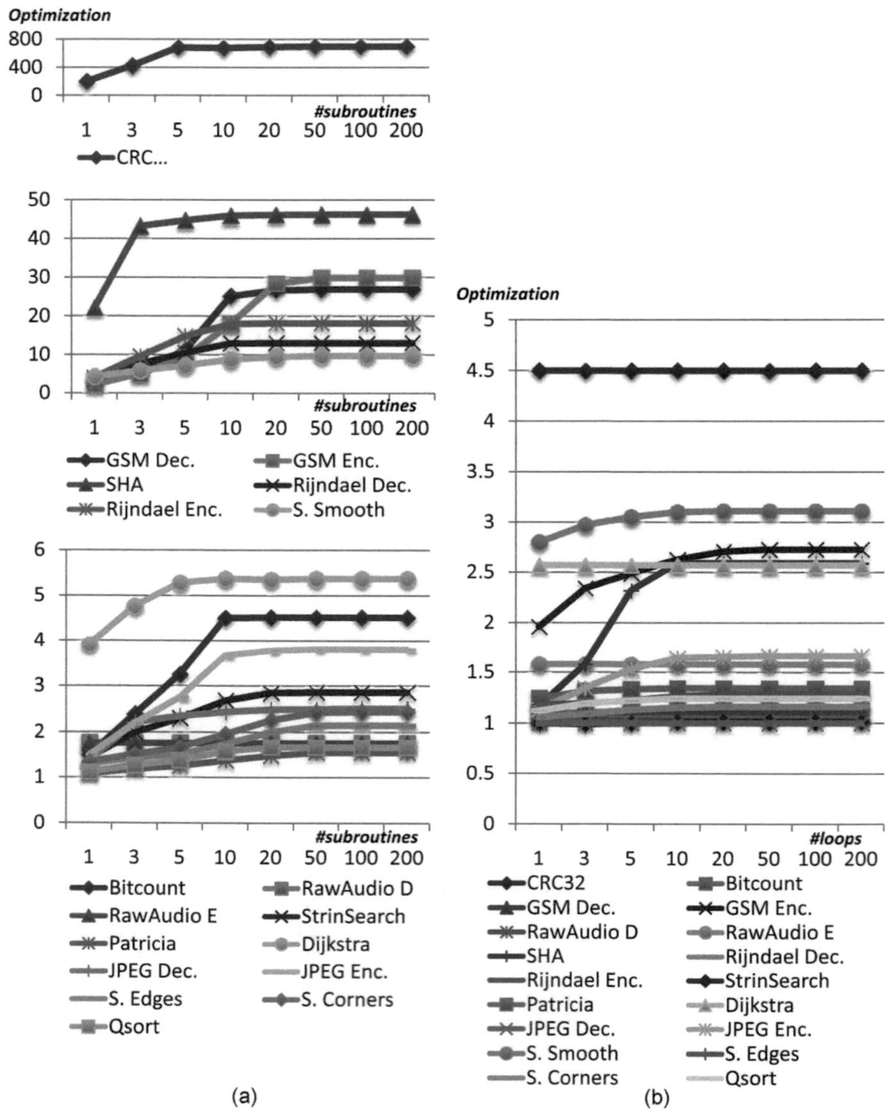

Fig. 2.18 Performance gains, but now considering 5 cycles per hot spot execution. (**a**) Subroutines and (**b**) loops

with the previous one, it can be observed that, although the algorithms that present performance speedups are the same, the speedup levels varies a lot.

Figure 2.19 presents the same analysis, but considering more pessimistic assumptions. Now, each hot spot would take 20 cycles to be executed. Although usually a reconfigurable unit would not take that long to compute one configuration, there are

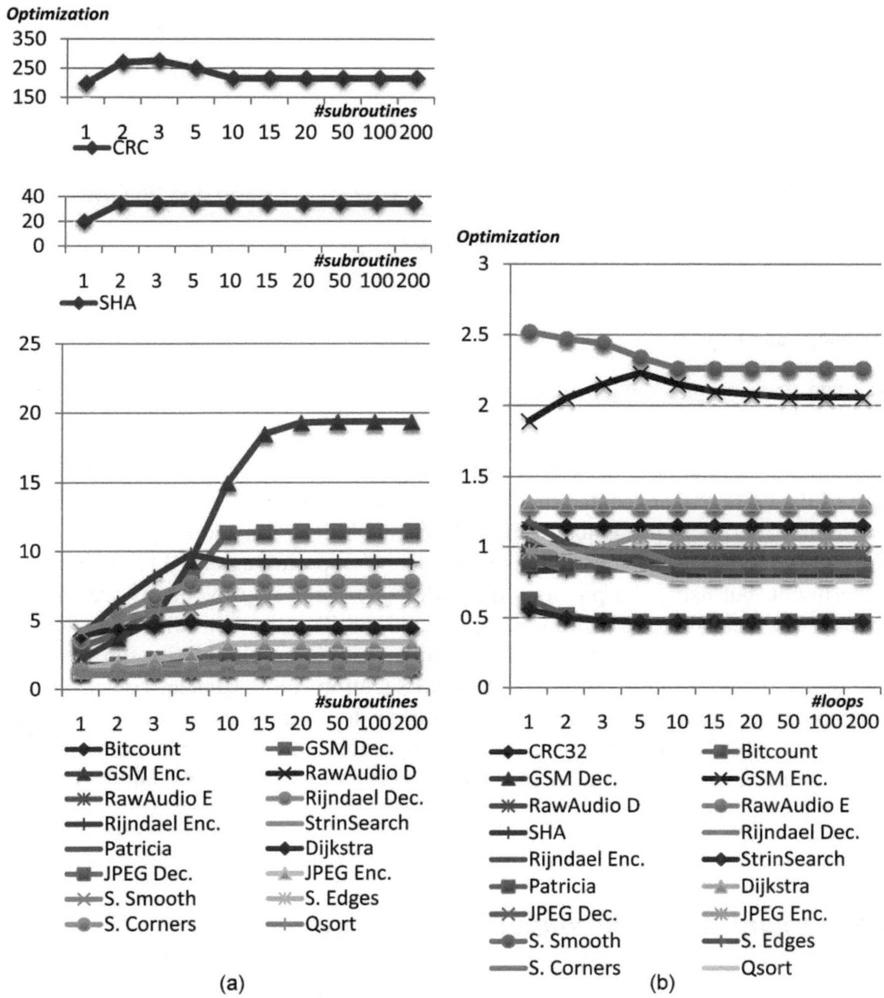

Fig. 2.19 Now considering 20 cycles per hot spot execution. (**a**) Subroutines and (**b**) loops

some exceptions, such as configurations with large code blocks, huge context sizes or those that have massive memory accesses. In the same figure, one can observe that some algorithms present even performance loss. This means that, depending on the way the reconfigurable logic is implemented and how the communication between the GPP and RU is done, some hot spots may not be worth to be executed on reconfigurable hardware.

Now, a different methodology is considered: it is assumed that enough reconfigurable hardware is available to support infinite configurations. This way, in opposite to the previous methodology, entire loops or subroutines could be optimized, regardless if all instructions inside them are executed in all iterations. Figure 2.20

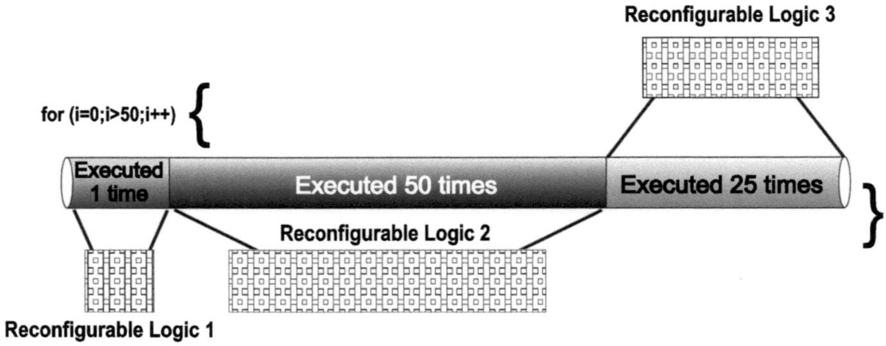

Fig. 2.20 Different pieces of reconfigurable logic are used to speed up the entire loop

illustrates this assumption. A different configuration would be available for each part of the code to be executed on reconfigurable hardware.

In the experiment presented in Figs. 2.21a and b, it is considered that the execution of each configuration would take 5 cycles. Comparing against Fig. 2.18 (same experiment, but using the previous methodology), huge improvements are shown, mainly when considering subroutine optimizations. This, in fact, reinforces the importance of using totally or partially dynamic reconfigurable architectures, which can adapt to the program behavior during execution. For instance, considering a partially reconfigurable architecture executing a loop: the part of the code that is always executed could remain in the reconfigurable unit during all the iterations, while sequences of code that are executed in certain time intervals could be configured when necessary.

2.6.3 Coarse Grain Reconfigurable Architectures

In this section, the potential of performance improvements considering that a coarse grain array is employed is analyzed. In these experiments, let us consider that the optimization will take place at the instruction level, and no speculative execution is supported. Therefore, the optimization is limited to basic block boundaries. Consequently, the level of optimization is directly proportional to the BBs execution rate (Fig. 2.14). For a determined basic block, the more it is executed, more performance boosts it represents for the overall acceleration.

Assuming that one configuration takes just one cycle to be executed, let us compare the instruction level optimization (representing the coarse grain architecture) against the subroutine level (representing the fine grain), which had previously shown more performance improvements than the loop level. When comparing the results in Fig. 2.22a to the ones in Fig. 2.17, one can observe that, for some algorithms, no matter how many basic blocks are optimized, the level of optimization will not reach the ones presented when executing only one subroutine in a fine grain

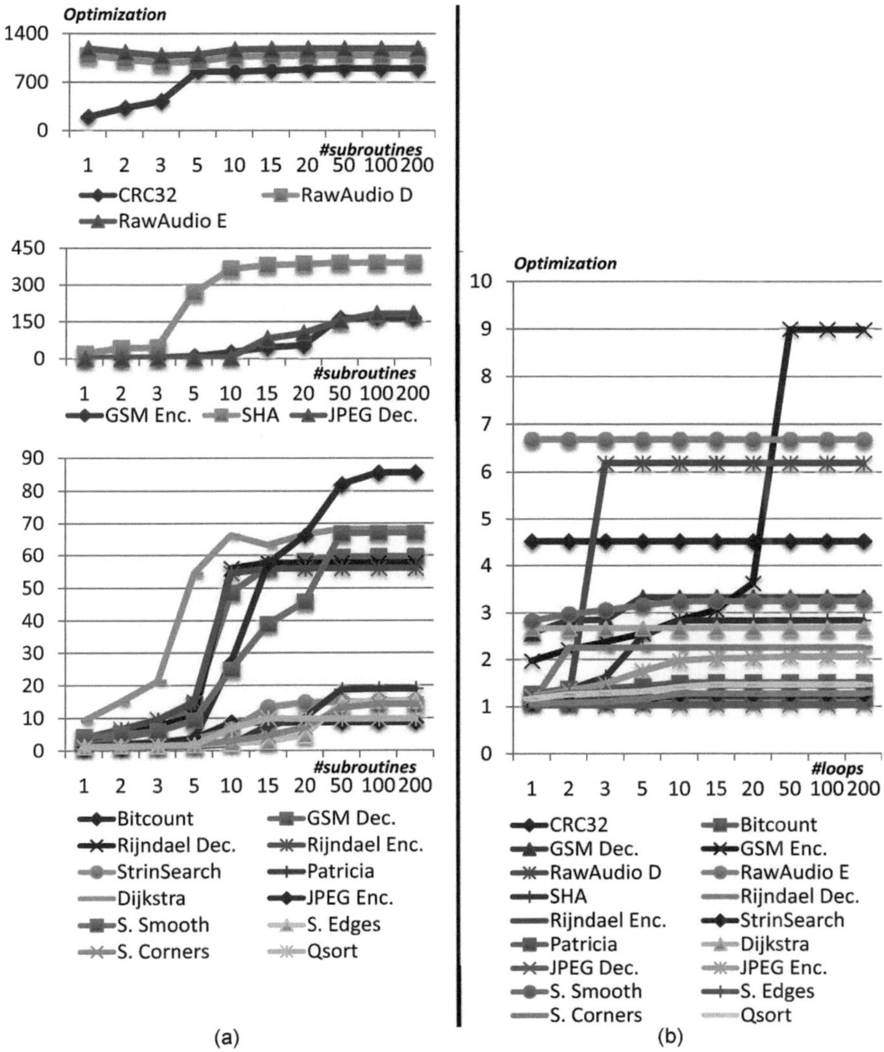

Fig. 2.21 Infinite configurations available for (**a**) subroutine optimization: each one would take 5 cycles to be executed. (**b**) The same, considering loops

system. However, for others, mainly the most complex ones, the level of optimization is almost the same for basic blocks or subroutines (they can be observed at the bottom of the figure). In the later case, using the instruction level optimization would be the best choice: it is easier and cheaper to implement 10 different configurations in coarse grain logic than 10 in a fine grain system, since much less connections and reconfigurable gates will be involved, with much less control circuits and associated delays.

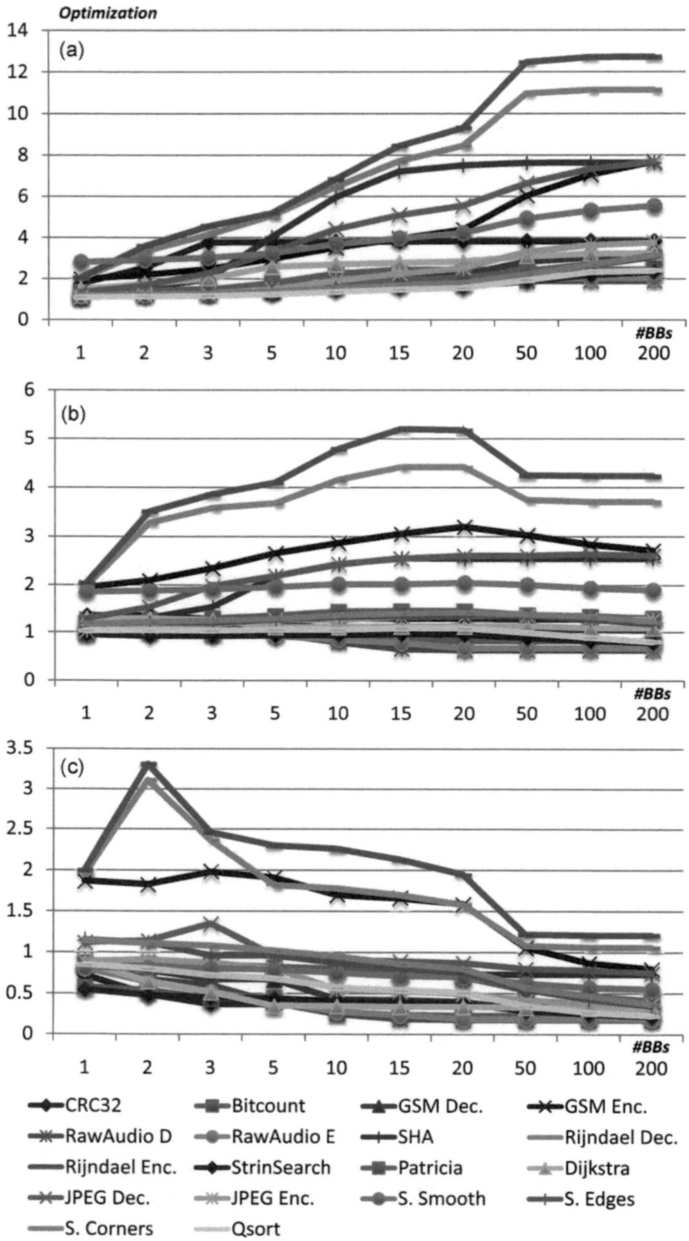

Fig. 2.22 Optimization at instruction-level with the basic block as limit. (**a**) 1 cycle, (**b**) 5 cycles, (**c**) 20 cycles per BB execution

When assuming that each configuration would take 5 cycles to be executed, there is a tradeoff between execution time and how complex the basic blocks are (in number of instructions, kind of operations, memory accesses etc.). This assumption is demonstrated in Fig. 2.22b: in the *Rinjdael* algorithms, the use of a coarse grain system is worth until a certain number of basic blocks being implemented on reconfigurable logic is reached. After that, there is performance loss. Considering 20 cycles per basic block execution (Fig. 2.22c), this situation becomes more evident. This shows that, as for fine grain reconfigurable architectures, there is a necessity of small reconfiguration and context loading times. These are easier to be achieved in coarse grain architectures though, since the size of each configuration is usually much smaller than fine grain ones.

Even though this coarse grain reconfigurable array does not demonstrate the same level of performance gains as fine grain reconfigurable systems show, more and different configurations would be available to be executed on this kind of system, considering that the memory size for keeping contexts would be smaller. In this case, the chances of optimization for applications that do not have very distinct kernels would increase when comparing against fine grain systems.

2.6.4 Comparing Both Granularities

Considering fixed applications, or data stream based ones, or yet those with long lifetime periods such as an MP3 player, fine grain reconfigurable systems may be a good choice. Some algorithms present huge potential for performance improvements, such as *CRC32*, *SHA* or *Dijkstra*. Only a small number of hot spots has to be optimized in these examples for them to present good acceleration results. This strategy, however, usually requires long development times, since a translation from the software code to a hardware description language amenable to synthesis must take place. Moreover, although there are tools that try to ease this task [33], their efficiency is quite limited, and several design iterations with human intervention are necessary.

Furthermore, it is important to point out a new industry trend: the number of different applications being executed on the systems is increasing and getting more heterogeneous. Considering the embedded systems field, some of the applications are not as datastream oriented as they used to be in the past. Applications with mixed (control and data flow) or pure control flow behaviors, where sometimes no distinct kernel for optimization can be found, are gaining popularity. Hence, for each application, different optimizations would be required. This, in consequence, would lead to an increase in the design cycle, since mapping code to reconfigurable logic usually involves some transformation, manual or using special languages or tool chains. The solution would be the employment of simpler coarse grain based reconfigurable architectures, even if they do not bring as much improvement as the fine grained approaches show.

The authors in [37] advocate in favor of coarse grain architectures. According to the authors, there are some reasons about why one should employ a coarse grain reconfigurable system, as follows:

- *Small configuration contexts*: Coarse grain reconfigurable units need few configuration bits, which are order of magnitude less than those required if FPGAs were used to implement the same operations. In the same way, a small amount of bits is necessary to establish the interconnections among the basic processing elements of coarse grain structures, since the interconnection wires are also configured at word level.
- *Reduced reconfiguration time*: Due to the previous statement, the reconfiguration time is reduced. This permits coarse-grain reconfigurable systems to be used in applications that demand multiple reconfigurations, even at run-time.
- *Reduced context memory size*: Being also a consequence of the first statement, the context memory size is also reduced. This allows the use of on-chips memories, which permits switching from one configuration to another with low configuration overhead.
- *High performance and low power consumption*: This stems from the hardwired implementation of coarse grained units and the optimal design of interconnections for the target domain.
- *Silicon area efficiency and reduced routing overhead*: Coarse grained units are optimally-designed hardwired units that are not built by combing a number of CLBs and interconnection wires, resulting in reduced routing overhead and better area utilization.

In contrast, according to the same authors, these are the main disadvantages of using a fine grain reconfigurable array such as the ones based on FPGA:

- *Low performance and high power consumption*: This happens mainly because word level modules need to be built by connecting a number of CLBs using a large number of programmable switches.
- *Large context and configuration time*: The configuration of CLBs and interconnections between them are performed at bit-level. This results in a large configuration context that has to be downloaded from the context memory, increasing configuration time, which may degrade performance when multiple and frequently-occurred reconfigurations are required.
- *Large context memory*: As a consequence of the previous statement, large reconfiguration contexts are produced, demanding a large context memory. Because of that, in many times the reconfiguration contexts are stored in external memories increasing even more the time and power necessary for reconfiguration.
- *Huge routing overhead and poor area utilization*: To build a word-level unit or datapath, a large number of CLBs must be interconnected, resulting in huge routing overhead and poor area utilization. In many times a great number of CLBs are used only for routing purposes and not for performing logic operations.

However, still according to the authors in [37], the development of universal coarse-grain architecture to be used in any application is an "unrealistic goal". This statement comes mainly from the fact that it is very hard to adapt the reconfigurable unit for a great number of different kernels, since the optimization is usually done at compile time. This way, even coarse grained architectures would be restricted to a specific domain. However, as it will be shown in the sequel, there are actually examples of reconfigurable accelerators that show excellent performance under a dynamic environment to optimize different kernels.

References

21. Athanas, P.M., Silverman, H.F.: Processor reconfiguration through instruction-set metamorphosis. Computer **26**(3), 11–18 (1993). doi:10.1109/2.204677
22. Barat, F., Lauwereins, R.: Reconfigurable instruction set processors: A survey. In: RSP'00: Proceedings of the 11th IEEE International Workshop on Rapid System Prototyping (RSP 2000), p. 168. IEEE Computer Society, Los Alamitos (2000)
23. Beck, A.C., Rutzig, M.B., Gaydadjiev, G., Carro, L.: Run-time adaptable architectures for heterogeneous behavior embedded systems. In: ARC'08: Proceedings of the 4th International Workshop on Reconfigurable Computing, pp. 111–124. Springer, Berlin/Heidelberg (2008)
24. Clark, N., Kudlur, M., Park, H., Mahlke, S., Flautner, K.: Application-specific processing on a general-purpose core via transparent instruction set customization. In: MICRO 37: Proceedings of the 37th Annual IEEE/ACM International Symposium on Microarchitecture, pp. 30–40. IEEE Computer Society, Los Alamitos (2004). doi:10.1109/MICRO.2004.5
25. Compton, K., Hauck, S.: Reconfigurable computing: a survey of systems and software. ACM Comput. Surv. **34**(2), 171–210 (2002). doi:10.1145/508352.508353
26. Guthaus, M.R., Ringenberg, J.S., Ernst, D., Austin, T.M., Mudge, T., Brown, R.B.: Mibench: a free, commercially representative embedded benchmark suite. In: WWC'01: Proceedings of the Workload Characterization, 2001. WWC-4. 2001 IEEE International Workshop, pp. 3–14. IEEE Computer Society, Los Alamitos (2001). doi:10.1109/WWC.2001.15
27. Hennessy, J.L., Patterson, D.A.: Computer Architecture: A Quantitative Approach, 4th edn. Morgan Kaufmann, San Mateo (2006)
28. Henning, J.L.: Spec cpu2000: Measuring cpu performance in the new millennium. Computer **33**(7), 28–35 (2000). doi:10.1109/2.869367
29. Hwu, W.M.W., Mahlke, S.A., Chen, W.Y., Chang, P.P., Warter, N.J., Bringmann, R.A., Quellette, R.G., Hank, R.E., Kiyohara, T., Haab, G.E., Holm, J.G., Lavery, D.M.: The superblock: an effective technique for vliw and superscalar compilation. In: Instruction-level Parallel Processors, pp. 234–253 (1995)
30. Jain, M.K., Balakrishnan, M., Kumar, A.: Asip design methodologies: Survey and issues. In: VLSID'01: Proceedings of the 14th International Conference on VLSI Design (VLSID'01), p. 76. IEEE Computer Society, Los Alamitos (2001)
31. Mahlke, S.A., Lin, D.C., Chen, W.Y., Hank, R.E., Bringmann, R.A.: Effective compiler support for predicated execution using the hyperblock. In: MICRO 25: Proceedings of the 25th Annual International Symposium on Microarchitecture, pp. 45–54. IEEE Computer Society, Los Alamitos (1992). doi:10.1145/144953.144998
32. Panainte, E.M., Bertels, K., Vassiliadis, S.: The Molen compiler for reconfigurable processors. ACM Trans. Embed. Comput. Syst. **6**(1), 6 (2007). doi:10.1145/1210268.1210274
33. Patel, S.J., Lumetta, S.S.: Replay: A hardware framework for dynamic optimization. IEEE Trans. Comput. **50**(6), 590–608 (2001). doi:10.1109/12.931895
34. Sima, D.: Decisive aspects in the evolution of microprocessors. Proc. IEEE **92**(12), 1896–1926 (2004)
35. Singh, H., Lee, M.H., Lu, G., Bagherzadeh, N., Kurdahi, F.J., Filho, E.M.C.: Morphosys: An integrated reconfigurable system for data-parallel and computation-intensive applications. IEEE Trans. Comput. **49**(5), 465–481 (2000). doi:10.1109/12.859540

36. Smith, M.J.S.: Application-Specific Integrated Circuits. Addison-Wesley, Reading (2008)
37. Theodoridis, G., Soudris, D., Vassiliadis, S.: A survey of coarse-grain reconfigurable architectures and cad tools. In: Fine- and Coarse-Grain Reconfigurable Computing, pp. 89–149. Springer, Dordrecht (2007). http://www.springerlink.com/content/j118u3m6m225q264/

Chapter 3
Deployment of Reconfigurable Systems

Abstract Following the discussion on reconfigurable systems, this chapter is dedicated to show several and different examples of architectures that have been used both in the academy and in the industry. They are presented according to the classification studied in the previous chapter. Also, a brief discussion on recent dataflow machines is done, since their structure is very similar to some of the reviewed reconfigurable systems. After that, one can find comparative tables summarizing the previously studied information. Then, an overview of the characteristics of the employed benchmark sets shows that all those proposed reconfigurable architectures fail on the very same and important aspect: they cannot cope with a large range of different applications in the same device, nor can they sustain binary compatibility. Therefore, it is made clear that some sort of dynamic optimization is necessary.

3.1 Introduction

In this chapter, some of the most known reconfigurable systems are presented. Each description is divided in sub-sections, based on the classification previously presented. A special sub-section is added, briefly discussing how the given architecture was evaluated and tested, which benchmark was used and presenting results regarding area and performance, when available. In the end of this chapter, a table summarizing the characteristics of reconfigurable systems is presented, as well as an analysis on the behavior of the employed benchmarks. In addition, there is a section briefly discussing recent dataflow architectures, since they are very similar to some of the reconfigurable architectures found nowadays. Even though there are recent surveys about the theme, both on coarse [61, 95] and fine grain [92] systems, they do not provide the comparison on different characteristics here presented.

A.C. Schneider Beck Fl., L. Carro, *Dynamic Reconfigurable Architectures and Transparent Optimization Techniques*,
DOI 10.1007/978-90-481-3913-2_3,

3.2 Examples of Reconfigurable Architectures

3.2.1 Chimaera

The Chimaera system [63, 64] was created with the claim that the custom computing units available at that time used to suffer with communication delays. Therefore, large chunks of the application code should be optimized to achieve reasonable performance improvements, so that one could compensate for the extra communication time.

3.2.1.1 RU Coupling

In order to decrease communication time, this was one of the first proposals of a reconfigurable system that actually works together with the host processor, as a tightly coupled unit, with direct access to its register file.

3.2.1.2 Reconfigurable System and Granularity

The main component of the system is the reconfigurable array, which consists of FPGA-like logic designed to support high-performance computations. Hence, its granularity is fine. The array is given direct read access to a subset of the registers in the processor (either by adding read connections to the host's register file, or by creating a shadow register file which contains copies of the values of those registers). The array is coupled to a MIPS R4000 processor. In the employed FPGA there are no state holding elements (such as flip-flops or latches) or pipeline stages inside it, making it totally combinational. The Chimaera system is illustrated in Fig. 3.1.

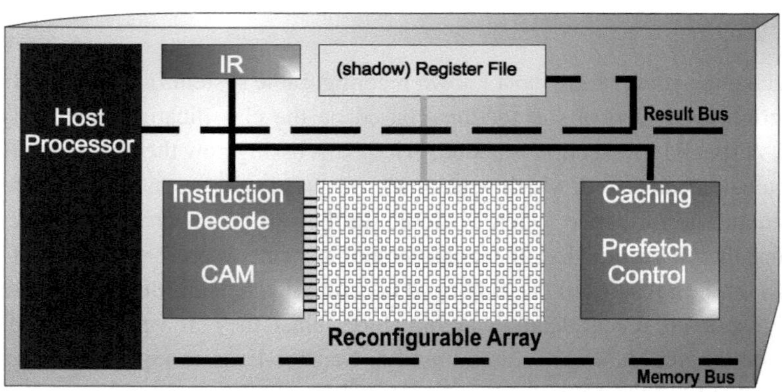

Fig. 3.1 Organization of the Chimaera system

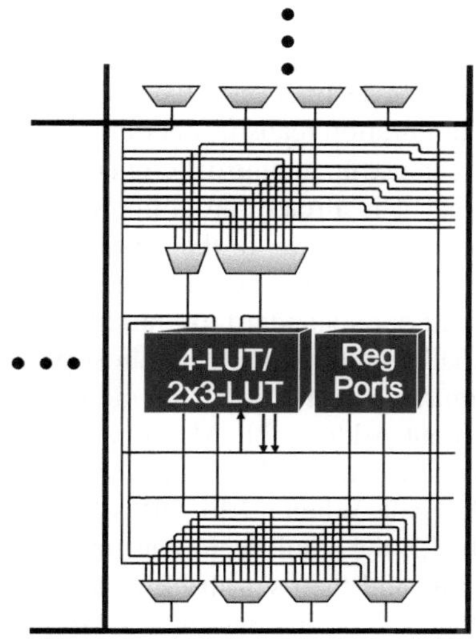

Fig. 3.2 Structure of the Chimaera reconfigurable array

The RU is divided in blocks (one is illustrated in Fig. 3.2). These blocks are connected to each other using multiplexers reassembling the behavior of a cross-bar. There are also special lines to connect blocks that are distant to each other. Each block is composed of register ports and a set of LUTs (one 4-input LUT, two 3-input LUTs or one 3-input LUT with a carry). The routing mechanism was also modified to allow partial reconfiguration at faster speeds. Another interesting aspect of this architecture is the downward flow of information and computation through the reconfigurable array. There is no way to send signals back to a higher row in the system, since the array behaves in a combinational-like fashion, without a feedback path.

3.2.1.3 Instruction Type, Reconfiguration and Execution

As part of the tasks of the host processor's decode logic, it should be determined whether the current instruction is a RFUOP opcode (name given to a reconfigurable instruction in the Chimaera). If that is true, a mechanism configures the RFU to produce the next result. In order to use the RFU, the application code includes calls to the RFU (using special instructions), and the corresponding RFU mappings are contained in the instruction segment of that application. Moreover, the system supports more than one instruction in the reconfigurable unit at the same time. Chimaera treats the reconfigurable logic not as a fixed resource, but rather as a cache for reconfigurable instructions. Instructions that have recently been executed, or those it can otherwise predict might be needed soon, are kept in the reconfigurable logic. If another instruction must be sent to the RFU, it needs to overwrite one or more currently loaded instructions. Consequently, it can be stated that the system supports partial reconfiguration.

The RFU call consists of the RFUOP opcode, indicating that an RFU instruction is being called, an ID operand that determines which specific instruction should be executed, and the destination register operand. The information from which registers an RFU configuration reads its operands is intrinsic in the instruction. A single RFU instruction can use up to nine different operands. If that instruction is already present (meaning that it is already programmed, or configured) in the RFU, the result of that instruction is written to the destination register during the instruction's write back cycle. In this way, the RFU calls act just like any other instruction, fitting into the processor's standard execution pipeline. If the requested instruction is not currently loaded into the RFU, the host processor is stalled while the RFU fetches the reconfigurable instruction from memory and properly reconfigures itself.

The Content Addressable Memory (CAM) of Fig. 3.1 determines which reconfigurable instructions are loaded in the array, where they are, and whether they are completed or not. When a RFUOP is found, and if the value in the CAM matches the RFUOP ID, the result from that row in the reconfigurable array is written onto the result bus, and then sent back to the register file, considering that the computation is done. If the instruction corresponding to the RFUOP ID is not present, the Caching/Prefetch control logic stalls the processor, and loads the proper RFU instruction from memory into the array. The caching logic also determines which parts of the reconfigurable array are overwritten by the instruction being loaded, and attempts to retain those RFU instructions most likely to be needed in the near future. Reconfiguration is done on a per-row basis, with one or more rows making up a given RFU instruction.

3.2.1.4 Code Analysis and Transformation

A C compiler, built over the GCC framework, was developed to transform groups of instructions to RFUOPs. It works by extracting subgraphs from a generated DFG, composed of instructions that can be executed in the array.

3.2.1.5 Evaluation

The system was evaluated with several benchmarks. In [63] it was reported that three different algorithms were used for the system validation: Compress/SPEC92, with a speedup of 1.11; Eqntott/SPEC92, presenting a speedup of 1.8; and Conway's Game of Life. By simply replacing the kernels with RFU instructions, it is possible to get a speedup of 2.06. With manual modifications via careful hand mapping of bit parallel optimization opportunities, a speedup of 160 times was achieved.

In [105], some applications of MediaBench [70] were evaluated: MPEG Encoder, G.721 encoder and decoder, ADPCM compression and decompression, Pegwit (public key encryption), as well as applications taken from the Honeywell benchmark: image compression and decompression. In [64], besides the ones cited above, other applications were also tested: DES encryption/decryption, Simple Gaussian Blur, RGB-Scale Conversion, RC5 encryption, Skeletonization Algorithm, DNA String Comparison and Compress.

3.2.2 GARP

3.2.2.1 RU Coupling

The GARP machine is a reconfigurable system attached to a MIPS II instruction set processor [46, 65] as a co-processor, so it can be classified as a loosely couple RU.

3.2.2.2 Granularity

The GARP system uses FPGA technology for the reconfigurable logic, so it is a fine grain reconfigurable array. However, because of that, it is necessary to overcome some obstacles, such as (according to the authors) [65]:

- It is claimed that FPGAs are rarely large enough to encode entire useful programs all at once. Consequently, smaller configurations handling different pieces of a program should be swapped in over time. However, the time needed for the swapping would be too expensive for any configuration to be used only briefly and discarded soon after. In real programs, several parts of code are not repeated often enough to be worth loading and executing into an FPGA.
- No circuit implemented with an FPGA can be as efficient as the same circuit in dedicated hardware. When compared to their counterparts in ordinary processors, standard functions like multiplications and floating-point operations are big and slow in an FPGA.
- Problems that are worth solving with FPGAs usually involve more data than can be kept in the FPGAs themselves. Furthermore, there is no standard model regarding the attachment of external memory to FPGAs. FPGA-based machines typically include ad hoc memory systems, designed specifically for the first application envisaged for the machine.
- Wide acceptance in the marketplace requires binary compatibility among a range of implementations. The current crop of FPGAs, on the other hand, must be reprogrammed for each new chip version, even within the same FPGA family.

3.2.2.3 Reconfigurable System

Garp's reconfigurable array is composed of entities called blocks (Fig. 3.3). One block in each row is a control block. The rest of the blocks in the array are logic blocks, which roughly correspond to the CLBs of the Xilinx 4000 series [76].

Considering the example given in [65], the Garp architecture fixes the number of column blocks at 24. The number of rows is implementation-specific, but can be expected to be at least 32. The basic quantum of data within the array is 2 bits. Logic blocks operate on values as 2-bit units, and all wires are arranged in pairs to transmit 2-bit quantities. This way, operations on 32-bit quantities generally require 16 logic blocks. Compared to typical FPGAs, Garp expends more hardware on accelerating

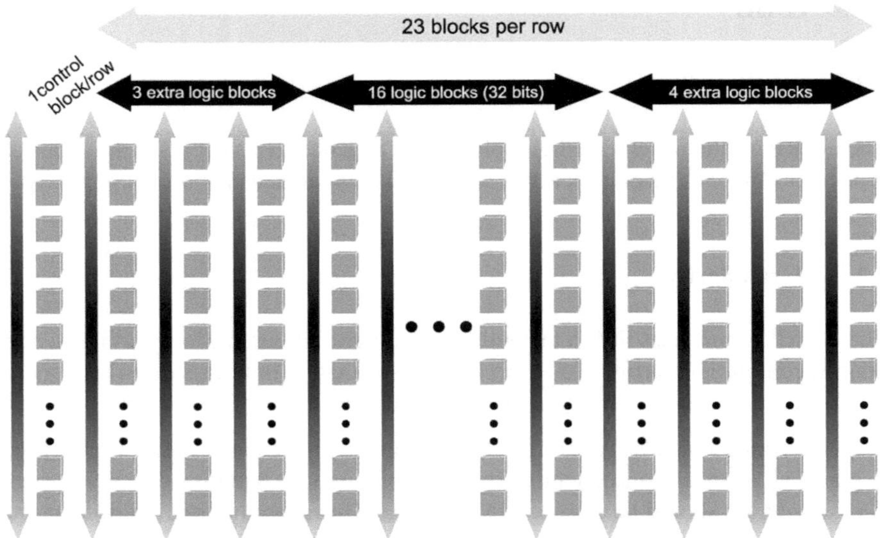

Fig. 3.3 A block of the GARP machine

operations like adds and variable shifts. The decision to make everything 2 bits wide is based on the assumption that a large fraction of most configurations will be taken up by multi-bit operations that are configured identically for each bit. By pairing up the bits, the size of configurations, the time required to load configurations, and the space taken up on the die to store them, are all reduced, at the cost of some loss of flexibility.

Rather than specify component delays as precise times that would change with each processor generation, delays in Garp are defined in terms of the sequences that can be fit within each array clock cycle. Only three sequences are allowed:

- short wire, simple function, short wire, simple function;
- long wire, any function not using the carry chain; or
- short wire, any function.

3.2.2.4 Instruction Type, Reconfiguration and Execution

With GARP, the loading and execution of configurations in the reconfigurable array is always under the control of a program running on the main processor. Several instructions have been added to the MIPS-II instruction set for this purpose, including ones for loading configurations, for copying data between the array and the processor registers, for manipulating the array, and for saving and restoring array state on context switches. The Garp reconfigurable hardware has direct access to the main memory system.

The use of the RU in Garp typically involves the following steps:

1. Load a configuration. This configuration can be found in a special cache designed particularly for holding the most used ones. If this is the case, this step would take a short time to be completed.
2. Copy the input data from the register bank to the reconfigurable array with the coprocessor move instructions.
3. Starts execution and wait until it is done.
4. After that, copy final results back to the register bank.

Each block in the array requires exactly 64 configuration bits (8 bytes) to specify the sources of inputs, the function of the block, and the wires driven with outputs. No configuration bits are needed for the array wires. A configuration of 32 rows requires approximately 6 KB. Assuming a 128-bit path to external memory, loading a full 32-row configuration takes 384 sequential memory accesses. At that time, a typical processor external bus might need 50 μs to complete the load.

Since not all useful configurations require the entire resources of the array, Garp allows partial array configurations. The smallest configuration is one row, and every configuration must fill exactly some number of contiguous rows. Two configurations can never be active at the same time, no matter how many array rows might be left unused by a small configuration.

3.2.2.5 Code Analysis and Transformation

The reconfigurable instructions are hand-coded and statically scheduled. A modified GCC-like design flow is used, using a pseudo language bonded together with the assembly generated from a C source.

3.2.2.6 Evaluation

Simulations were performed in order to gather results for Garp, since at that time no actual hardware implementation existed. It was compared against a Sun Ultra-SPARC 1/170, a 4-way superscalar 64-bit processor with 16 kB each of on-chip instruction and data caches. Performance estimates can be observed in Table 3.1. In Fig. 3.4 one can observe the area estimates of a hypothetical implementation in hardware of this reconfigurable system. It would be implemented in a 0.5 μm, 4-metal-layer process in a die size of 17.5×17.8 mm^2. It is also compared with the same UltraSPARC (Fig. 3.5).

Table 3.1 Performance estimates for GARP machine, compared to the SPARC [46]

Function	Data Size	Speedup
Image median filter	640 x 480 pixels	43
DES (ECB mode)	1 Megabyte	18.7
Image dithering	640 x 480 pixels	17.0
strlen	1,024 chars	14.2
strlen	16 chars	1.84
Sort	2 Megabytes	2.2

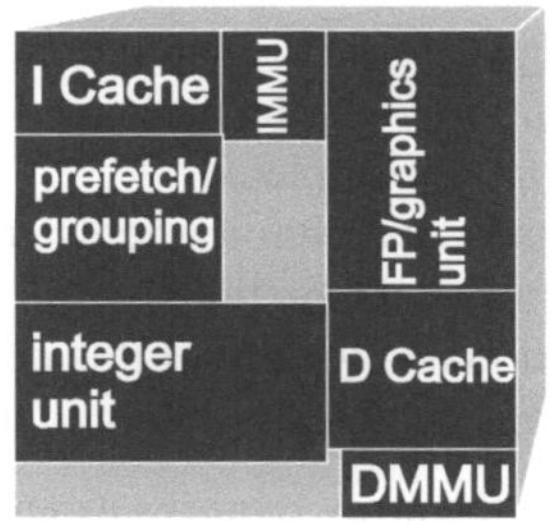

Fig. 3.4 Area estimates for the GARP System [65]

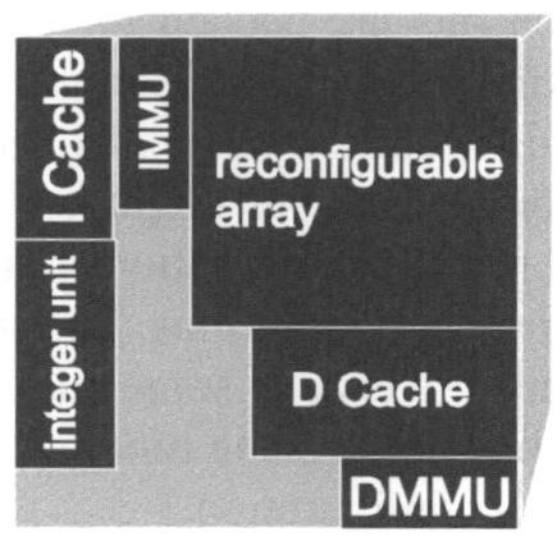

Fig. 3.5 Area estimates for the Ultrasparc processor [65]

3.2.3 REMARC

REMARC comes from "Reconfigurable Multimedia Array Coprocessor" [82, 83]. It is a reconfigurable unit, coupled to a MIPS II ISA based RISC machine. As the name states, REMARC was specifically designed to speed up multimedia applications.

3.2.3.1 RU Coupling

The MIPS ISA can support up to four coprocessors. In this case, coprocessor 0 is already used for memory management and exception handling, coprocessor 1 is used for a floating point unit. Then, REMARC operates as coprocessor 2. REMARC is a loosely coupled reconfigurable architecture.

3.2.3.2 Reconfigurable System and Granularity

A coarse grain reconfigurable system was employed, because, according to the authors [83], fine grain FPGA based reconfigurable architectures have the following drawbacks:

- The small width of the programmable logic blocks results in large area and delay overheads to implement wider datapaths, such as 8 or 16 bits long.
- FPGAs are slower when compared to a custom integrated circuit and have lower logic density.

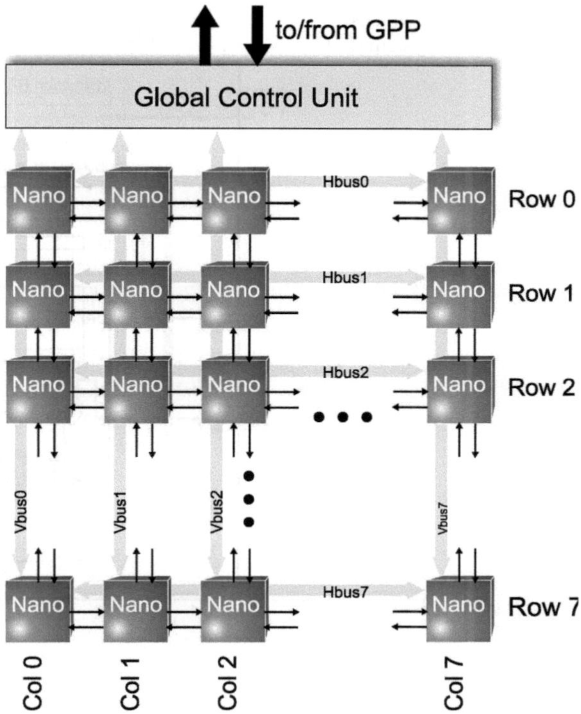

Fig. 3.6 General overview of the REMARC reconfigurable system

REMARC consists of an 8 × 8 array of nano processors and a global control unit. A nano processor can communicate to the four adjacent ones through dedicated connections and to the processors in the same row and the same column through the 32-bit Horizontal Bus (HBUS) and the 32-bit Vertical Bus (VBUS), respectively. A general overview of the system can be observed in Fig. 3.6. The global control unit is responsible for controlling the execution of the nano processors (the communication is done using the VBUSes) as well as for the data transfer between them and the GPP.

The nano processor consists of a 32-entry instruction RAM, a 16-bit ALU, a 16-bit entry data RAM, an instruction register (IR), eight 16-bit data registers (DR), four 16-bit data input registers (DIR), and a 16-bit data output register (DOR). The DOR registers are used to receive data from the four adjacent nano processors (up, down, left, and right) through dedicated connections (DINU, DIND, DINL, and DINR). The DOR register data can also be used as source data for ALU operations or data inputs of a DIR register. These local connections provide high bandwidth pathways within the processor array. The 16-bit ALU can execute 30 different instructions. The nano processor is demonstrated in Fig. 3.7. Taking advantage of the availability of SIMD instructions in multimedia applications, one instruction can be used for the whole set of nano processors that belong to a column or to a row.

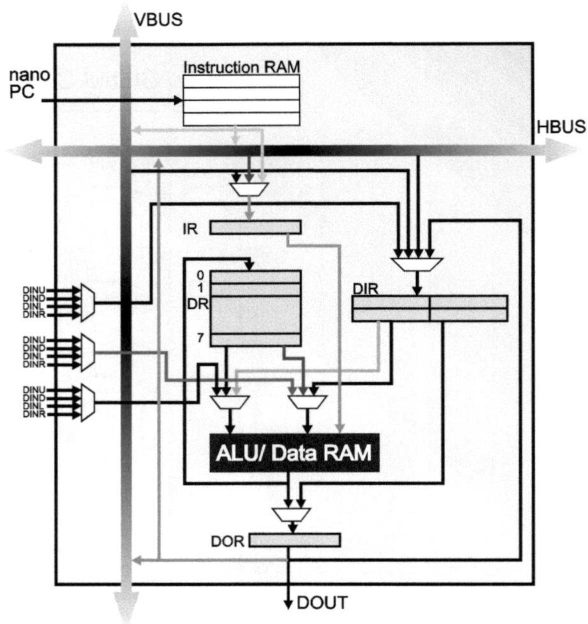

Fig. 3.7 One nano processor in the REMARC system [83]

3.2.3.3 Instruction Type, Reconfiguration and Execution

The nano processors do not have Program Counters (PCs) by themselves. The global control unit generates the PC value (nano PC) for all nano processors every cycle. All nano processors use the same nano PC and execute the instruction indexed by it. However, each nano processor has its own nano instruction RAM. Therefore, each nano processor can operate differently according to the nano instructions stored in this local RAM. This makes it possible to achieve a limited form of Multiple Instruction Stream, Multiple Data Stream (MIMD) operation in the processor array. At this point, according to the authors, REMARC can be regarded as a VLIW processor in which each instruction consists of 64 operations.

As already stated before, the global control unit controls the nano processors and the transfer of data between them and the main processor. It includes a 1024-entry instruction RAM (global instruction RAM), data registers, and control registers. These registers can be accessed by the main processor directly with the following instructions: move from/to coprocessor or load/store from/to coprocessor. Moreover, the GPP has the role of controlling the RU. GPP loads the operators, starts RU execution and writes back the results. The MIPS ISA was extended to support such instructions.

3.2.3.4 Code Analysis and Transformation

The reconfigurable instructions are programmed in the special REMARC assembly language, and can be added to a regular C code using GCC.

3.2.3.5 Evaluation

Remarc executes MPEG2 decoding, optimizing two kernels: IDCT and MC. It also executes MPEG2 encoding and DES. A high-level simulation of the system demonstrated speedups ranging from a factor of 2.3 to 21.2 in the aforementioned applications.

3.2.4 Rapid

The goal of RaPiD [52, 53, 56] was to compile regular computations like those found in DSP applications into both an application-specific datapath, and the program for controlling that datapath.

3.2.4.1 RU Coupling, Reconfigurable System and Granularity

RaPiD is a standalone (no GPP works together with it), coarse-grain architecture that allows dynamic construction of deeply pipelined computational datapaths from a mix of ALUs, multipliers, registers and local memories. The ALUs perform the usual logical and arithmetic operations on signed or unsigned fixed-point 16-bit data. A general overview of RaPiD-I system is shown in Fig. 3.8. RaPiD is composed only of a configurable datapath and a sequencer. This way, the reconfigurable datapath is responsible for executing the whole program.

The datapath is shown in more details in Fig. 3.9. As can be observed, the components of the datapath are arranged logically in a linear array. The functional units

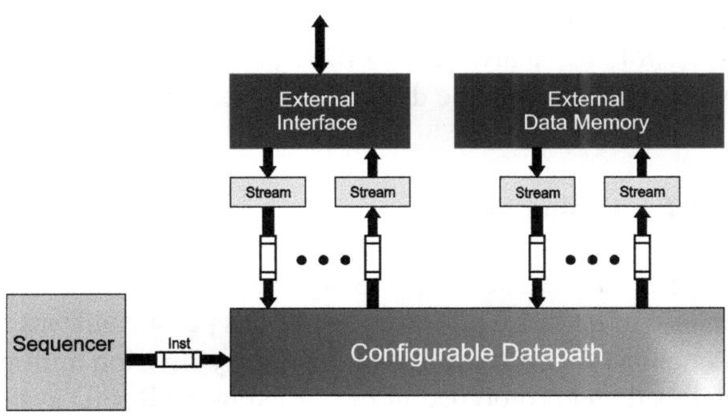

Fig. 3.8 RaPiD-I system [53]

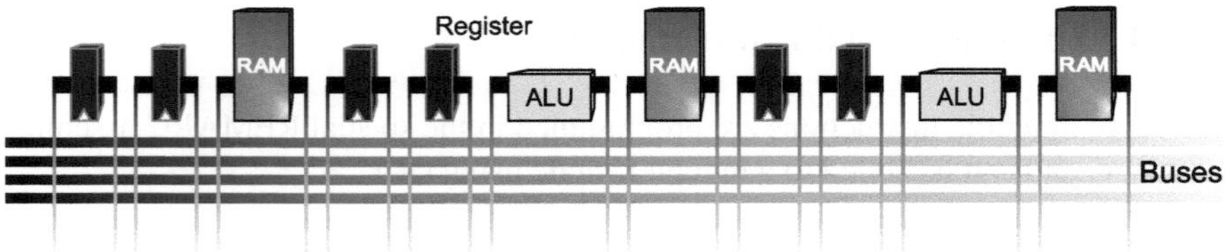

Fig. 3.9 RaPiD-I cell [53]

are interconnected using a set of ten segmented buses that run the length of the datapath. Each input of the functional units is attached to a multiplexer that is configured to select one of the buses. The output of each functional unit is attached to a demultiplexer comprised of tristate drivers, driving the signals to one of the buses. Each output driver can be configured independently, which allows an output to fan out to several buses, or none at all if the functional unit is not being used. As case-study, an array was developed [52]. It was comprised of 16 cells, each containing one multiplier/shifter, three embedded memories (used for temporary variables, constant tables etc.), three ALUs and six registers.

3.2.4.2 Instruction Type, Reconfiguration and Execution

RaPiD is programmed for a particular application by first mapping the computation onto a datapath pipeline. The control signals are divided into static control signals (also called hard control) provided by the configuration memory, and dynamic control (soft control) which must be provided on every cycle. The static programming bits are used to construct the datapath structure, while the dynamic programming bits are used to schedule the operations of the computation onto the datapath over time. The controller is programmed to generate the information needed to produce the dynamic programming bits.

3.2.4.3 Code Analysis and Transformation

Programs for the RaPiD architecture are written in a modified C-like language, called RaPiD-C, which is an explicit data parallel language. The compiler is responsible for analyzing RaPiD-C programs to deliver the static datapath circuit, as well as to generate the dynamic control signals.

3.2.4.4 Evaluation

In [53] RaPiD executes two algorithms: FIR filter and Matrix multiply. A performance of up to 1.6 billion of operations per second was achieved. In [57], a Multiple-Antenna OFDM (Orthogonal frequency-division multiplexing) Application was developed and prototyped in FPGA using the RaPiD system.

3.2.5 Piperench (1999)

The main novelty of Piperench [51, 54, 59, 60] was the so-called "pipelined reconfiguration". It means that a given kernel is broken into pieces, and these pieces can be reconfigured and executed on demand. This way, the parts of a given kernel are multiplexed in time and space into the reconfigurable logic. This process, called virtualization, will be better explained next.

3.2.5.1 RU Coupling

In its current implementation, PipeRench can be classified as an attached processor, being loosely coupled.

3.2.5.2 Reconfigurable System and Granularity

Piperench is a coarse grain array. It interesting to repeat some of the reasons that motivated the authors to build this architecture without using an FPGA:

- Logic granularity: It is claimed that FPGAs are designed for logic replacement. The granularity of the functional units is optimized to replace random logic, not to perform multimedia computations.
- Configuration time: The time to load a configuration in the fabric ranges from hundreds of microseconds to hundreds of milliseconds. For FPGAs to improve processing speed over that of a general-purpose processor, they must amortize this start-up latency over huge data sets, limiting their applicability.
- Forward compatibility: FPGAs require redesign or recompilation to benefit from future chip generations.
- Hard constraints: FPGAs can implement only kernels of a fixed and relatively small size. This size restriction makes compilation difficult and causes large, unpredictable discontinuities between kernel size and performance.
- Compilation time: A kernel's synthesis, placement, and routing design phases take hundreds of seconds, taking longer than the compilation of the same kernel for a general-purpose processor.

Figure 3.10 presents a general overview of the PipeRench architecture. A set of physical pipeline stages are called stripes. Each stripe has an interconnection network and a set of Processing Elements (PEs).

In Fig. 3.11 one can observe a more detailed view of a PE. Each PE contains an arithmetic logic unit and a pass register file. Each ALU in the PE contains lookup tables (LUTs) and extra circuitry for carry chains, zero detection, and so on. Designers can implement combinational logic using a set of NB-bit-wide ALUs. They can also cascade the carry lines of these ALUs to construct wider ALUs by chaining them together via the interconnection network, so it is possible to build complex combinational functions.

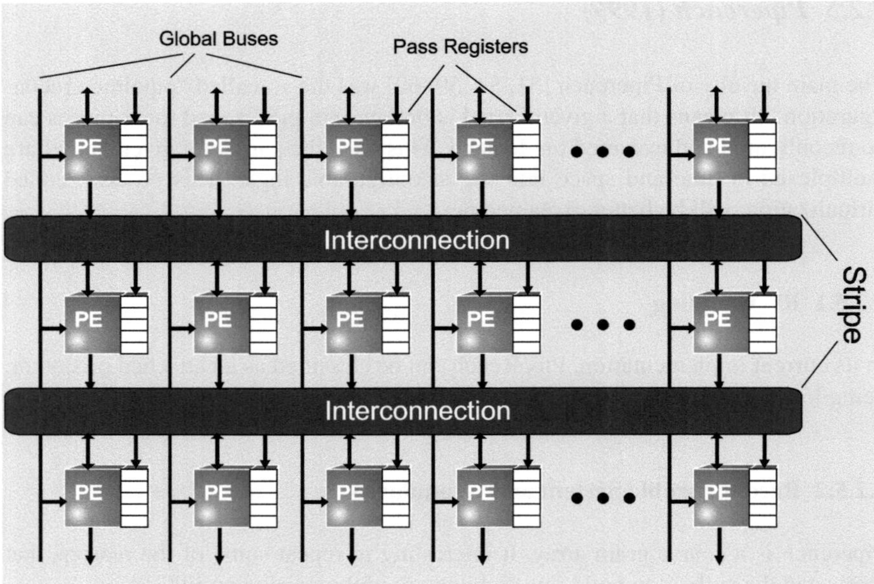

Fig. 3.10 General overview of the Piperench structure

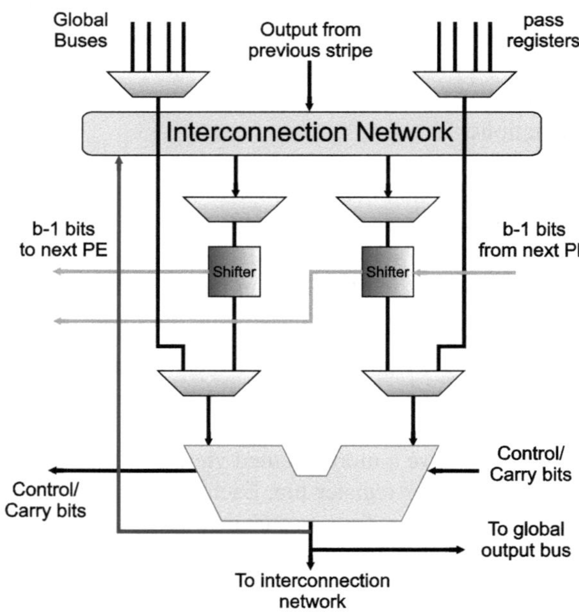

Fig. 3.11 Detailed view of the Process Element and its connections

Through the interconnection network, PEs can access operands from registered outputs of the previous stripe, as well as registered or unregistered outputs of the other PEs in the same stripe. The pass register file provides a pipelined interconnection from a PE in one stripe to the corresponding PE in subsequent stripes. A program can write the ALU's output to any of the P registers in the pass register file. If the ALU does not write to a particular register, that register's value will come from the value in the previous stripe's corresponding pass register. For data values to move laterally within a stripe, they must use the interconnection network. In each stripe, the interconnection network accepts inputs from each PE in that stripe, plus one of the register values from the previous stripe. Moreover, a barrel shifter in each PE shifts its inputs $B - 1$ bits to the left. Thus, PipeRench can handle the data alignments necessary for word-based arithmetic. The PEs can also access global I/O buses. These buses are necessary because an application's pipeline stages may physically reside in any of the fabric's stripes. Inputs to and outputs from the application must use a global bus to get to their destination. Because of hardware virtualization constraints, the buses cannot be used to connect consecutive stripes.

3.2.5.3 Instruction Type, Reconfiguration and Execution

The basic Piperench principles of reconfiguration and execution are based on the virtualization process. Figure 3.12 illustrates how virtualization works. In the upper part of this same figure (Fig. 3.12a), it is demonstrated an application which was divided in 5 different pipeline stages, taking the total of 8 cycles to be configured

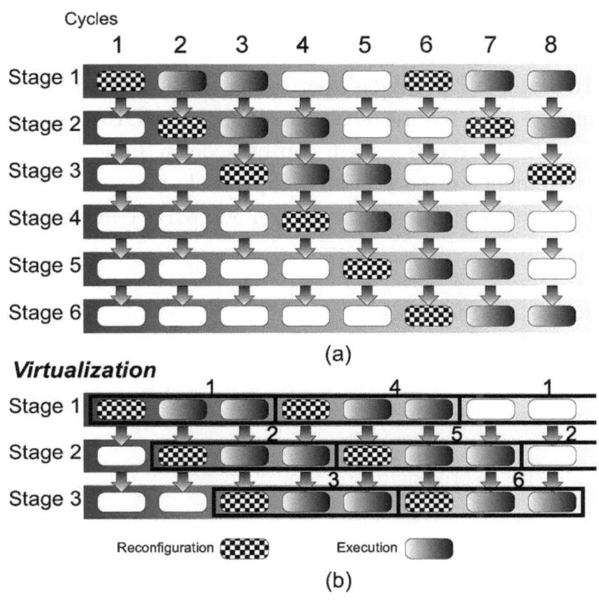

Fig. 3.12 The virtualization process, technique used by Piperench. (**a**) Normal execution. (**b**) With virtualization

and executed (each stage can be configured and used independently of each other), representing the regular operation. Figure 3.12b shows how this application can fit in the reconfigurable hardware after virtualization: just 3 stages of the equivalent pipeline stages presented before are necessary. The pipeline stages are reconfigured on demand, according to the kernel needs. Note that the virtual stage 1 is used to execute the equivalent of stages 1 and 4 of the original operation. This is feasible because it is done in different periods of time. Since some stages are configured while others are executed, partial reconfiguration does not decrease performance. Consequently, it is possible to execute the same piece of software taking the same time, but with a smaller area overhead. The ALU operation is static during the time a particular virtual stripe resides in a physical stripe.

3.2.5.4 Code Analysis and Transformation

The process of code generation uses a parameterized compiler. The compiler begins by reading a description of the architecture. This description includes the number of PEs per stripe, each PE's bit width, the number of pass registers per PE, the interconnection topology, PE delay characteristics, and so on. The source language is a dataflow intermediate language. After parsing, the compiler inlines all modules, performs loop unrolling, and generates the program.

3.2.5.5 Evaluation

To evaluate PipeRench's performance, the authors have chosen the following benchmarks: ATR, Cordic, DCT, DCT-2D, FIR, IDEA, Nqueens, Over, PopCount. Results for the Piperench system can be seen in Table 3.2. It shows the performance improvements of the 100 MHz Piperench, built as a 128-bit-wide fabric having 8-bits PEs with 8 registers each, over a 300 MHz Ultrasparc II.

Table 3.2 Performance improvements over a 300-MHz Ultrasparc II

Benchmark	Speedup
ATR	**189.7**
Cordic	**15.5**
DCT	**11.3**
DCT-2D	**12**
FIR	**63.3**
IDEA	**42.4**
Nqueens	**26**
Over	**57.1**
PopCount	**29**

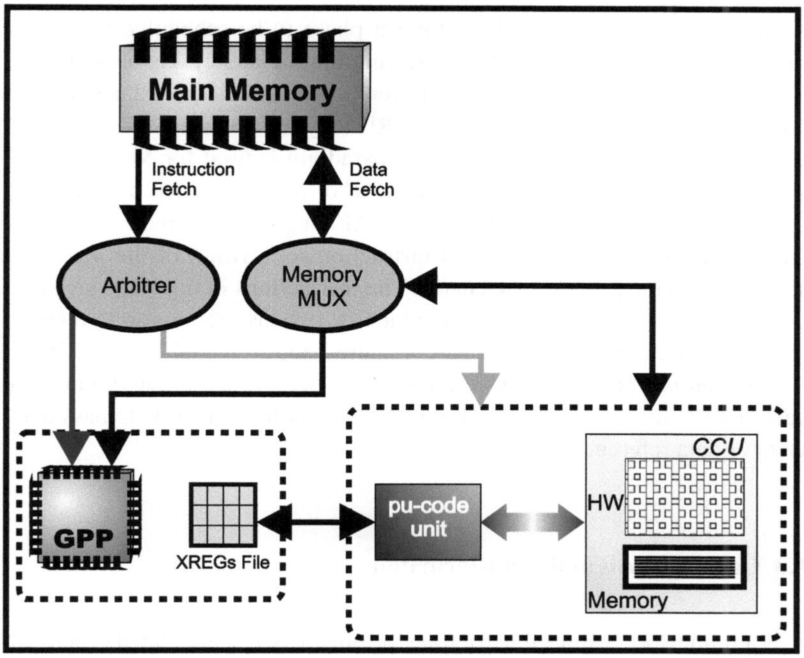

Fig. 3.13 A general overview of the Molen System

3.2.6 Molen

3.2.6.1 RU Coupling, Reconfigurable System and Granularity

The Molen processor [98, 99] is a reconfigurable system based on an FPGA with a loosely coupled reconfigurable array. The two main components in the Molen organization are depicted in Fig. 3.13. More precisely, they are the Core Processor, which is a GPP, and the Reconfigurable Unit (RU). The Arbiter issues instructions to both processors; and data transfers are controlled by the Memory MUX. The reconfigurable unit (RU), in turn, is subdivided into the $\rho\mu$-code unit and the Custom Computing Unit (CCU). The CCU is implemented in reconfigurable hardware, e.g., a field-programmable gate array (FPGA), and memory. The application code runs on the GPP except for the accelerated parts implemented on the CCU used to speed up the overall program execution. Exchange of data between the main and the reconfigurable processors is performed via the exchange registers (XREGs).

3.2.6.2 Instruction Type, Reconfiguration and Execution

The reconfigurable processor operation is divided into two distinct phases: *set* and *execute*. In the set phase, the CCU is configured to perform the targeted operations. Subsequently, in the execute phase, the actual execution of the operations takes place. Such decoupling allows the set phase to be scheduled well ahead of the execute phase, thereby hiding the reconfiguration latency. As no actual execution is performed in the set phase, it can even be scheduled upward across the code boundary in the instructions preceding the RU targeted code [84].

A sequential consistency programming paradigm is used for Molen [97]. It requires only a one-time architectural extension of a few instructions that supports a large user reconfigurable operation space. Although the complete ISA extension comprises 8 instructions, the minimal instruction set (ΠISA) of the $\rho\mu$-code unit is enough to provide a working scenario. The instructions in this class are: *set*, *execute*, *movtx* and *movfx*. By implementing the first two instructions (*set/execute*), any suitable CCU implementation can be configured and executed in the CCU space. The *movtx* and *movfx* instructions are needed to provide the input/output interface between the RU targeted code and the remaining application code to pass data, parameters or data references.

3.2.6.3 Code Analysis and Transformation

There is a framework responsible for transforming, from an annotated C application, the binary to be executed on the PowerPC processor, with the instructions responsible for handling the reconfiguration and execution processes.

3.2.6.4 Evaluation

In [99], Molen was evaluated with the MPEG2 encoder/decoder. The most time consuming operations among SAD (sum of absolute difference), 2D-DCT (two dimensional discrete cosine transform), and 2D-IDCT (two dimensional inverse DCT) were optimized. Table 3.3 demonstrates the impact of implementing these kernels as Molen hardware when comparing against a PowerPC processor without it, executing different video sequences. Columns labeled with "theory" present the theoretically achievable maximum speed up. Columns labeled with "impl." contain the speedups with respect to the considered Molen implementation.

Table 3.3 Molen Speed ups [99]

	MPEG2 encoder		*MPEG2 decoder*	
	theory	*impl.*	*theory*	*impl.*
carphone	2.85	2.64	2.02	1.94
claire	2.99	2.80	1.60	1.56
container	3.12	2.96	1.68	1.63
tennis	3.37	3.18	1.68	1.65

3.2.7 Morphosys

3.2.7.1 RU Coupling, Reconfigurable System and Granularity

MorphoSys [71, 88, 89] is a coarse-grain reconfigurable system. It was designed to operate on 8 or 16-bit data. As can be observed in Fig. 3.14, the system comprises of a RU, a modified RISC GPP (TinyRISC) and a memory interface unit with a high bandwidth. Both processors and RU are resident on the same chip. However, being different components, the RU is loosely coupled. The on-chip DMA controller enables fast data transfers between main memory and Frame Buffer, which is a memory interface, working similarly to a data cache.

The RU is organized as an array of Reconfigurable Cells (RCs) to facilitate SIMD operations. The reconfigurable array is composed of 64 Reconfigurable Cells (organized as an 8×8 array). The RC is coarse-grain, with one ALU and multiplier, one shifter, two multiplexers and a register file. Besides, there is the output and feedback registers for data exchange. The RC is illustrated in Fig. 3.15.

The RC is configured through a 32-bit word, saved in the Context Register, found in each RC. These words are stored in a special memory, called context memory. The Context Memory provides context words to the RC Array in each cycle of execution. Besides configuring the RC, they are also responsible for programming the interconnection network.

The interconnection network can be seen in Fig. 3.16. It is comprised of three hierarchical levels. The lowest level is a 2-D mesh, providing neighbor connectiv-

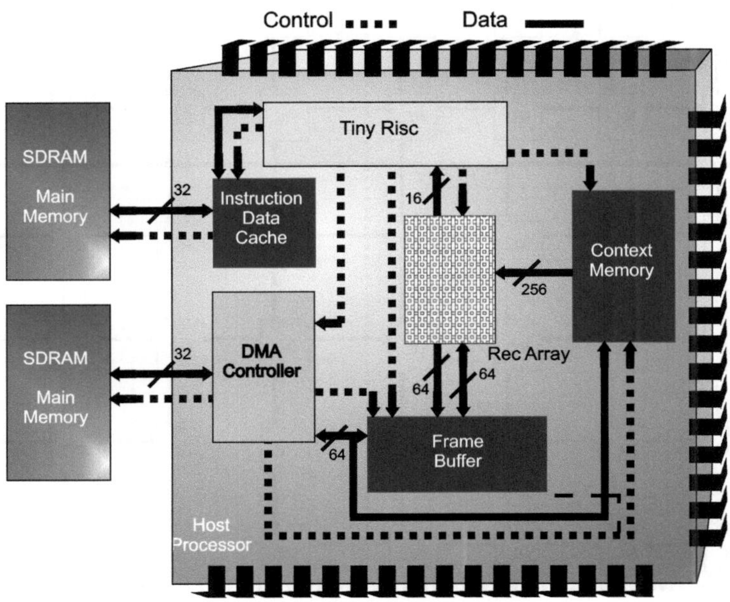

Fig. 3.14 Morphosys Organization

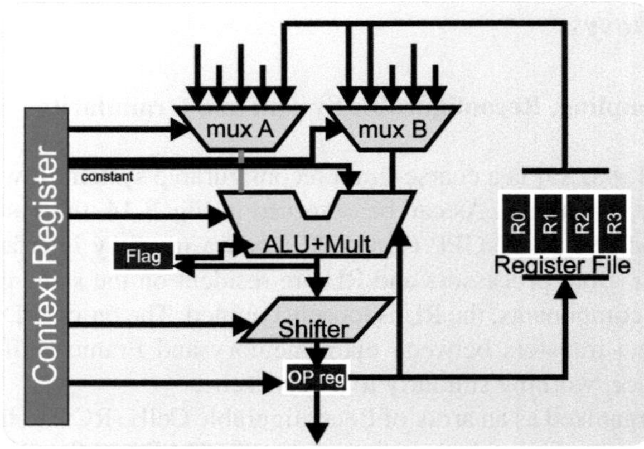

Fig. 3.15 The Morphosys Reconfigurable Cell

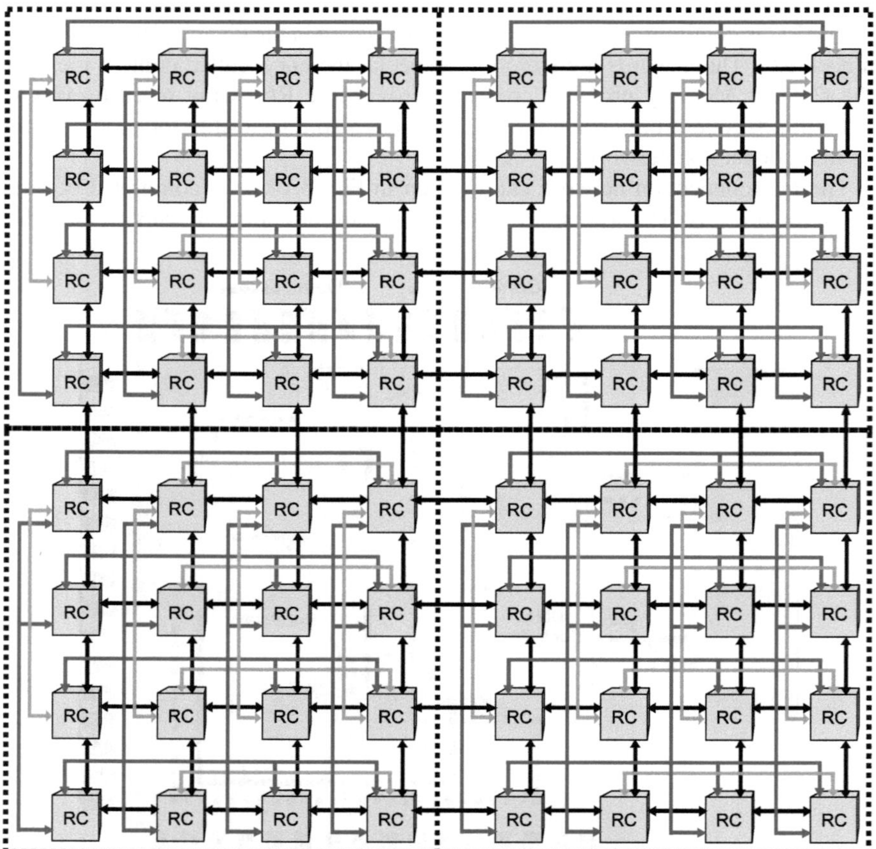

Fig. 3.16 The Morphosys Interconnection Network

ity. Then, the "Intra-quadrant" level divides the array in four equal parts, each one composed of 16 RCs (4×4). The array has 4 quadrants. Within each quadrant, a given RC can access the output of any other cell in its same row or column. The last level is the "Inter-quadrant", composed of buses between adjacent quadrants, running across rows and columns. These buses are responsible for the communication between RCs that are placed in different quadrants, but in the same row or column.

3.2.7.2 Instruction Type, Reconfiguration and Execution

The TinyRISC is responsible for controlling the RC array by the addition of special instructions to its ISA. These instructions perform the following functions: data transfer between main memory (SDRAM) and Frame Buffer; loading of context words from main memory into the internal Context Memory; and execution control of the RC Array. Context data may be loaded into a non-active part of Context Memory without interrupting RC Array operation. The Context Memory can store up to 32 configurations.

The following steps are necessary for using the RU in the Morphosys system: load from external memory context words into the Context Memory; Load computation data from the Frame Buffer from external memory; execute the computation in the RC. At the same time, it is possible to load more data to the Frame Buffer. This way, one is allowed to overlap actual computations with data transfers.

3.2.7.3 Code Analysis and Transformation

A graphical user interface was developed, called mView. It takes some user inputs for each application (specification of operations and data sources/destinations for each RC) and generates assembly code for the MorphoSys RC Array. This interface can also be used to simulate the system. A prototype compiler that compiles hybrid code for MorphoSys (from C source code) has been developed using the SUIF compiler environment [101]. The compilation is done after partitioning the code between the TinyRISC processor and the RC Array.

3.2.7.4 Evaluation

The system was evaluated with the following programs [71, 89]: Video Compression (MPEG) Motion Estimation for MPEG, Discrete Cosine Transform (DCT) for MPEG and Data Encryption/Decryption (IDEA Algorithm) and Automatic Target Recognition (ATR). They were compared to different ASICs, GPPs and reconfigurable systems.

3.2.8 ADRES

3.2.8.1 RU Coupling

ADRES [79] is a coarse-grained reconfigurable matrix tightly coupled to a Very Long Instruction Word (VLIW) processor.

3.2.8.2 Reconfigurable System, Granularity, Instruction Type, Reconfiguration and Execution

The ADRES, instead of being a fixed reconfigurable system, is generated from a template, based on a XML description language. It is used to define things such communication topology, employed operation set, resource allocation etc. The organization of the matrix is also not fixed. For instance, the functional units and register files can be organized in diverse forms: two functional units could share only one register file and so on. An example of a generated architecture can be observed in Fig. 3.17.

In ADRES, although both are part of the same physical entity, the VLIW processor and the reconfigurable matrix are virtually two different components. Because of this model, they can share resources. Some of the components of the VLIW processor are reused in the reconfigurable matrix, as can be observed in Fig. 3.17. It is important to point out that, although in a minor quantity, the functional units of the

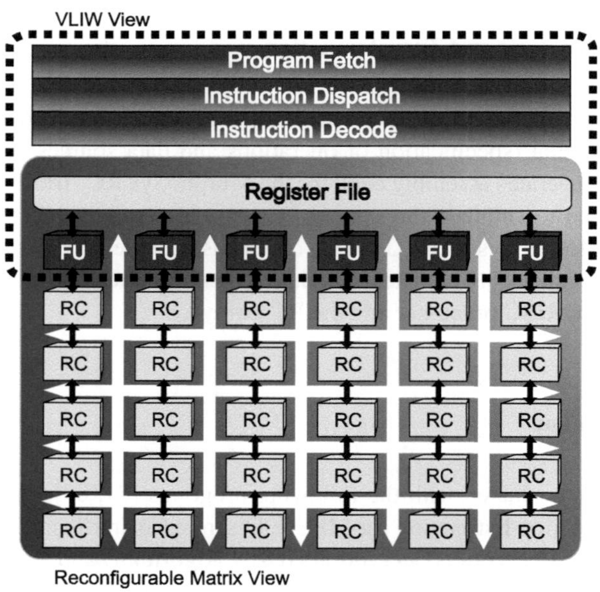

Fig. 3.17 An example of an ADRES based reconfigurable system

VLIW processor can execute more operations when comparing to those placed in the reconfigurable matrix.

Data communication is done by sharing both memory and register file. According to the authors [79], this helps mapping high-level languages such as C to the ADRES architecture more easily. The register file and memory cannot be simultaneously shared in order to avoid synchronization and data integrity issues.

3.2.8.3 Code Analysis and Transformation

A framework was developed [80], together with a scheduling algorithm to be used with the ADRES architecture. The IMPACT compiler is employed as frontend [50] to parse the source code (in the C language). After optimization and analysis, it generates an intermediate representation, called *lcode*. After that, the compiler tries to find loops that can be optimized by the reconfigurable matrix. For the rest of the code that is mapped to the VLIW processor, regular techniques are used in order to find the best possible ILP. Finally, these two separated parts are merged together to be executed on the system. The scheduler takes into account the fact that there are shared resources between the VLIW processor and the reconfigurable unit and communication between both parts is also considered.

3.2.8.4 Evaluation

The testbench used in [79] is composed of 4 different programs. They are typical digital signal processing and multimedia applications. They are derived from C reference code of Texas Instruments DSP benchmarks and MediaBench [70]. IDCT is an 8×8 inverse discrete cosine transformation; *adpcm-dec* refers to an ADPCM decoder; *mat mul* is a matrix multiplication; and a FIR filter. Table 3.4 shows the speedups presented by ADRES. In [78] an H.264/AVC decoder was implemented to be executed in ADRES, achieving a speed up from 1.3 to 1.9 times, depending on the input bitstream.

Table 3.4 ADRES Speed-ups over the standalone VLIW processor

Application	*Total ops (ADRES)*	*Total cycles (ADRES)*	*Total ops (VLIW)*	*Total cycles (VLIW)*	*Speed-up*
IDCT	211 676	6 097	181 853	38 794	6.4
Adpcm_dec	8 150 329	594 676	5 760 116	1 895 088	3.2
Matrix mult	20 010 518	1 001 308	13 876 972	2 811 011	2.8
FIR	69 126	3 010	91 774	18 111	6.0

3.2.9 *Concise*

3.2.9.1 RU Coupling and Granularity

The Concise system [68] is a tightly coupled, fine grain reconfigurable unit, based on a CPLD.

3.2.9.2 Reconfigurable System, Instruction Type, Reconfiguration and Execution

The RFU was included into the pipeline of a very simple RISC processor. This RFU is driven by specific RFU instructions, generated using a smart compiler that will be described in the next subsection. This way, the ISA of the processor was extended to support these new RFU instructions. The extra functional unit based on CPLD works in parallel to the ALU of the processor. The RU does not support partial reconfiguration.

The main principle of Concise is to minimize latency caused by reconfiguration. For that, multiple reconfigurable instructions are encoded within one configuration, so the number of times the RU must be reconfigured is reduced. Figure 3.18 demonstrates Concise. The RU (that follows a RISC traditional register-to-register format) is composed of the opcode, two sources and one destination registers. Two fields are specific for the reconfiguration mechanism. The DEC field goes directly to the CPLD, indicating which function it should do, while the CONF field indicates which configuration the CPLD should use. This way, each configuration of the

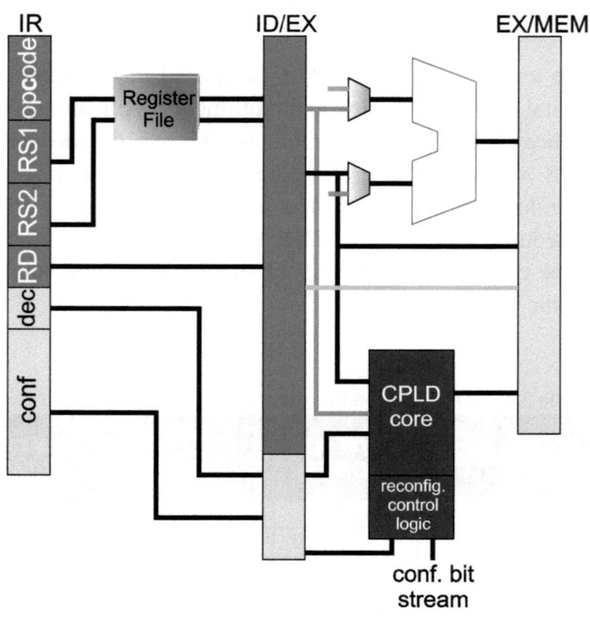

Fig. 3.18 The Concise System

CPLD should have a decoder to understand the DEC field. One advantage claimed by the authors is that using this approach the latency time of switching from one reconfigurable instruction to another would be reduced. This way, it would be possible to get good results even when two custom instructions are near to each other.

3.2.9.3 Code Analysis and Transformation

A smart compiler was developed in order to try to find optimal results in terms of area and performance for each custom instruction and to hide as much as possible the steps of RFU instruction generation from the programmer.

First, the source code, usually written in C, is processed by the compiler front-end, generating an intermediate code, which is represented by a DFG. Then, the framework looks for hot spot candidates, using profile data. Just arithmetic and logic operations are considered. These candidates are grouped in clusters, that will be transformed later in a CPLD configuration. The grouping process follows a certain criteria. For instance, candidates that are found in the same loop are usually grouped together; while hot spots with lower logic complexity are grouped in bigger clusters.

These clusters are sent to a translator, which transforms these clusters in HDL. The decode logic is added, so the different custom instructions that are placed in the same cluster can be executed independently. After that, the HDL is synthesized to hardware. Timing and fitting information is sent back to the cluster detection and selection mechanism. Considering this data, a cluster can be rearranged or even discarded. This cycle keeps going until a solution that can be considered satisfactory is found. The DFG segments that were transformed to reconfigurable instructions are labeled in the DFG. Final code is then generated, with register allocation, instruction scheduling and the assembly code. This assembly is sent to another assembler, which recognizes the labels that correspond to reconfigurable instructions. Finally, the netlist generated in the synthesis is combined with the assembly, so the final executable is ready.

3.2.9.4 Evaluation

In [68], the architecture is evaluated with encryption applications: DES (Data Encryption Standard, which is a block cipher) and A5 (stream cipher). Speedups reached 40% when compared to the standalone RISC processor.

3.2.10 PACT-XPP

The main purpose of PACT-XPP (eXtreme Processing Platform) [44] is to execute data-stream software in the array, using runtime and self reconfiguration mechanisms.

3.2.10.1 RU Coupling

PACT-XPP is a standalone system, meaning that it does not work with any other GPP.

3.2.10.2 Reconfigurable System, Granularity, Instruction Type

The reconfigurable unit of XPP is based on a hierarchical array. This array is coarse grain, composed of PAEs (Processing Array Elements), which communicate to each other using the communication network. The PAE is composed of PAE objects. They can be memory, ALU etc. For example, the PAE that is based on ALU is composed of registers for receiving and sending data, besides the ALU itself for performing the computations. The structure of a PAE is shown in Fig. 3.19. According to the authors [44] any desired functionality can be added to the XPP architecture. PAE objects are called of self-synchronizing type, because their operation starts soon after data input packets are available for them. In the same way, their results are forwarded as soon as they are ready, so they can be consumed by other PAEs.

Rectangular blocks of PAEs compose a PAC (Processing Array Cluster). A device based on the XPP architecture has one or more PACs. The example in Fig. 3.20 contains four PACs. Each PAC is attached to a Configuration Manager (CM), responsible for the configuration of that block.

As already stated, the PAE communicate to each other using a packet oriented network. There is a hardware protocol to ensure data will not be lost, so no explicit scheduling of operation is necessary. Two kinds of packets can be sent through this network: data and event packets. Data packets have a fixed bit width, depending on the implementation of the device. Event packets usually are a few bits long. They

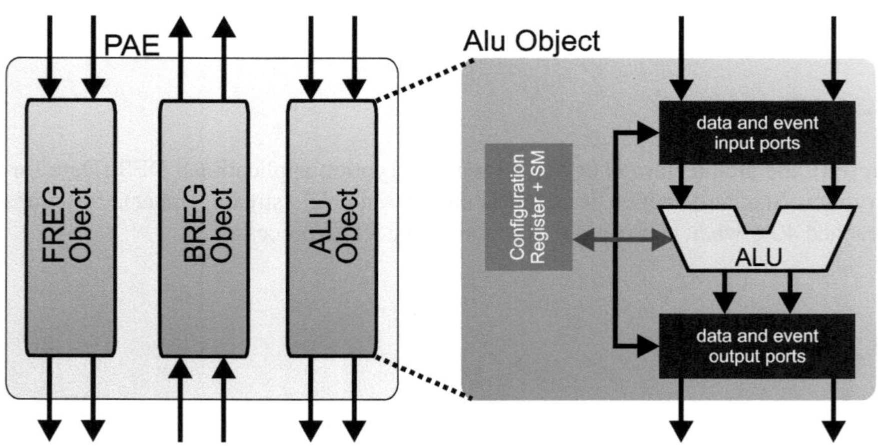

Fig. 3.19 An example of a PAE and a and ALU-object in PACT-XPP

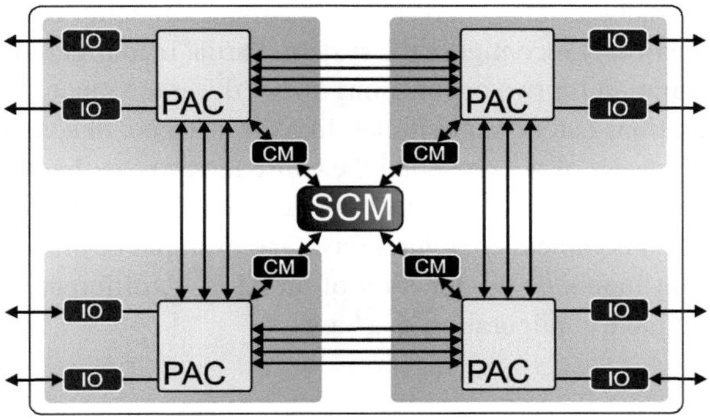

Fig. 3.20 A set of Process Array Clusters (PACs)

transmit state information to the network, such as the ALU state. Hence, the transmission of conditional computations, depending on the state of previously executed data, is possible. Event packets can also trigger a self reconfiguration, meaning that part of the array can be configured because of some computation that has occurred internally. These signals can also be used to control the data streams.

3.2.10.3 Reconfiguration and Execution

The configuration control is not centralized, but rather distributed: several CMs are responsible for that. The CM is basically a state machine with an internal RAM, used for configuration caching. There is one CM for each PAC. This way, it is possible to configure some PACs at the same time while others are executing an operation. Hence, entire applications can be executed in different parts of the array simultaneously.

Devices with several packs contain additional Configuration Managers, so a hierarchal tree of CMs is formed (Fig. 3.20). The root is called SCM (Supervising CM). The SCM has an external interface that connects it with the configuration memory. The interface is composed of address and data buses, plus control signals. The reconfiguration can be activated in two ways: externally or by special events that were originated inside the array. The authors call this approach of self-reconfiguring design [44].

Each PAE holds an internal state. For instance, a PAE can be configured or free, so the control system can figure if a PAE can be reconfigured at a given moment or not. This way, a PAE cannot be reconfigured while still being used for another application. Furthermore, while executing the computation of a specific application, the structure of the array (configuration) used for that application remains static: connections or operators are not changed. The XPP supports pre fetching of configurations in order to hide configuration latency, besides the possibility of fetching configuration while PAEs are still executing another.

Besides the possibility of configuring or removing an entire configuration, it is also possible to partially reconfigure the system. Partial reconfiguration can be used when the configuration for two applications do not differ too much. In certain cases, distinct configurations can be very similar, like for instance, in adaptive filters. The authors call the process of not changing the entire application, but rather just a part of it, of differential configuration. According to them, this kind of configuration is more effective than complete configurations. One example of this operation is just changing constant inputs or the function of an ALU. A differential configuration always has a complete configuration as a base.

3.2.10.4 Code Analysis and Transformation

To map the application onto the reconfigurable system, a structural language with reconfiguration primitives was developed. It is called NML (Native Mapping Language). The NML allows the access of all hardware features to the programmer. NML is similar to structural HDLs, such as VHDL. PAE objects are explicitly allocated. Furthermore, they can also be placed according to the programmer desires. Connections between components can also be specified. There is a modified C compiler, called XPP-VC [47, 48]. It has the role of translating C functions to NML modules. This C compiler is restricted by a subset of C language and the use of a specific library. The programmer can mix both approaches. This way, it is possible to hand code the most critical hot spots in NML, in order to achieve maximum performance, while using the C compiler for the rest of code.

3.2.10.5 Evaluation

The case studied in [44] was evaluated with the algorithms presented in Table 3.5. This same table also shows the performance of each program in terms of ops/cycle and gigaops/sec.

Table 3.5 Pact-XPP: Performance

Application	OPS/Cycle	GigaOPs/sec
Integer FIR	256	38.4
Complex FIR	290	43.5
Median Filter	38	5.7
Coordinate Transf.	36	5.4
Matrix Mult.	64	9.6
MPEG4 Inner Loop	31	4.7
FFT8 Kernel	32	4.8

3.2.11 RAW

3.2.11.1 RU Coupling

RAW, or Reconfigurable Architecture Workstation [94, 100], is a standalone and totally independent system, not working with any GPP.

3.2.11.2 Reconfigurable System and Granularity

RAW is divided in 16 parts, called tiles (Fig. 3.21). Each tile is composed of routers; a MIPS like processor, with a 8 stages pipeline; a floating point unit with a four stages pipeline; 32 Kb of data and 96 Kb of instruction caches. Each tile was designed to take one clock cycle to compute data. RAW does not use buses, rather, it uses a switched interconnection network. According to the authors, the fact that memory is distributed across the tiles eliminates the memory bandwidth bottleneck, providing significantly lower latencies to each memory module.

The switches can be programmed both dynamically and statically. The later means that the switches are scheduled before execution starts and are maintained during the whole program lifecycle. The static router is pipelined, and controls two routing crossbars (so there are two physical networks to interconnect the tiles with this purpose). Each router can send values to different places: north, east, south, west neighbors; to the GPP and to the other crossbar. To support dynamic routing, a pair of wormhole oriented networks was added to the architecture.

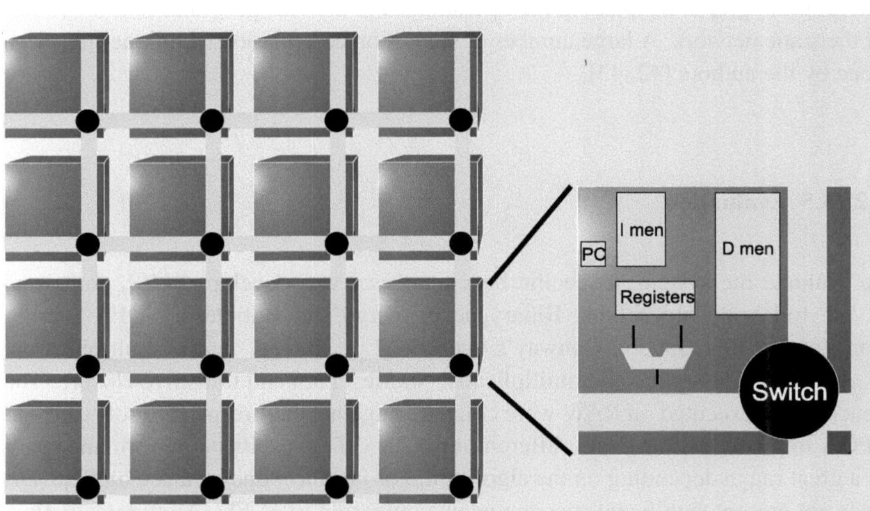

Fig. 3.21 RAW Organization

3.2.11.3 Instruction Type, Reconfiguration and Execution

Each tile runs a single thread, having its own program counter, separated from each other. As commented before, the switches can be programmed with static communication, so the compiler can be responsible for threads communication, meaning that synchronization issues can be amortized.

An operating system was developed. It is responsible for dynamic scheduling of processes and context switches, as any conventional GPP. Each process can use one or more tiles. The placement of these processes in the architecture can be variable, meaning that each process does not necessarily need to be always executed on the same place. This fact is hidden to the user, though. This way, the placement of tasks inside the network is virtualized, in the sense that the system adapts itself at run time to the running process. The OS always allocates a rectangular-shaped number of tiles to each process.

3.2.11.4 Code Analysis and Transformation

The architecture is visible to the programmer and compiler, so one can program the routing, having direct access to the data transfer mechanism. The main role of the specific compiler is to take a single or multi threaded programs written in any high level programming language, and map it to the hardware. A specific compiler (C and Fortran) was developed, called RawCC [72], and it is responsible for partioning the program graph, for placing the operations and also for programming the routes of the static network. A large number of studies on compilation techniques has been done by the authors [42, 43].

3.2.11.5 Evaluation

To evaluate the system, a specific benchmark set was developed [40], composed of the following algorithms: Binary heap, merge and Bubblesort, DES encryption, Integer FFT, Jacobi, Conway's Game of Life, Integer matrix multiplication, N queens, single-source and multiplicative shortest path and transitive closure. The benchmarks executed on RAW were compared against their respective software and FPGA implementations, with different input data. The results on performance vary in a great range depending on the algorithm. For instance, one instance of N queens does not present significant speedup when comparing to its FPGA implementation, while the Conway's Game of Life being executed on RAW achieves a speedup factor of 1758 when comparing to its software implementation.

3.2.12 *Onechip*

3.2.12.1 RU Coupling

The Onechip [49, 104] is a reconfigurable system tightly coupled to a MIPS processor, working as another functional unit in the execution stage.

3.2.12.2 Reconfigurable System and Granularity

The system can be composed of one or more reconfiguration units. These units are used in parallel with the basic functional units found in the processor. They are called PFU (Programmable Functional Units). While the basic functional unit is build in fixed logic and responsible for the regular MIPS operations, such as arithmetical and logic ones, PFUs can implement any application specific function, according to a given application. The PFUs use the processor components such as registers and memory interface as if they were the basic functional unit. Because of new instructions are added and none are modified, it is claimed that binary compatibility is maintained, in the sense that old code still can be executed after the Onechip architecture is coupled to the MIPS processor.

One prototype of the Onechip System was developed in FPGA technology, as can be observed in Fig. 3.22. The reconfigurable logic takes much more area then the fixed resources. Hence, the main processor was placed in the center, immersed in the middle of the reconfigurable area, so the distance between the processor and the reconfigurable units is balanced.

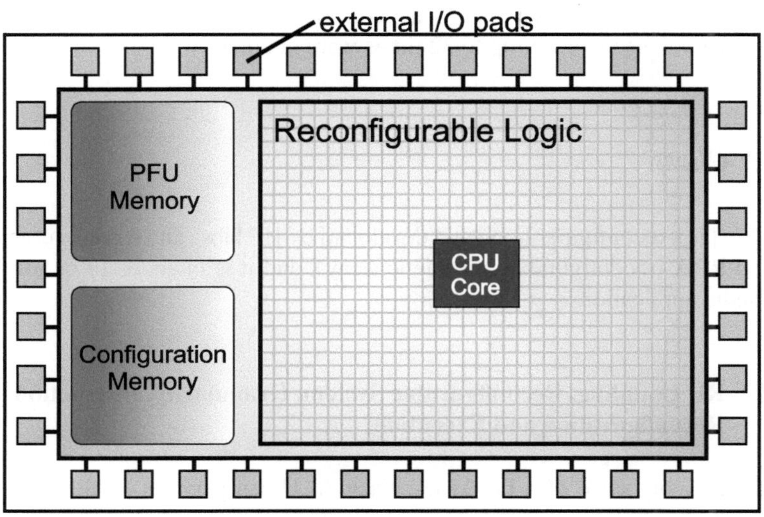

Fig. 3.22 General overview of the Onechip implementation

3.2.12.3 Code Analysis and Transformation

In [49] it is stated that code annotation is used in order to help the assembler to identify those instructions that are related to the reconfigurable system, so it could be simulated. However, the designer is responsible for both code annotation and hot spots identification.

3.2.12.4 Instruction Type, Reconfiguration and Execution

The unit is accessed using the new added instructions. In the example given in [104], a PFU was programmed to behave as a universal asynchronous receiver and transmitter. This would be executed in Onechip with the following instructions:

```
URTR REG (UART read instr.)
SW mem-loc, REG (STORE WORD instr.)
```

The REG is any of the general purpose registers available in the MIPS processor, and mem-loc is a memory pointer. As can be observed, PFU instructions share the same resources with the rest of the processor.

3.2.12.5 Evaluation

Onechip was evaluated with one and two dimensional versions of DCT [104]. Moreover, the system was coupled to different processors. In [49], four applications were tested: JPEG Image compression, ADPCM Audio coding, PEGWIT Data encryption and MPEG2 Video encoding. When considering the unit coupled to an out-of-order processor, speedups of up to 32 times were obtained.

3.2.13 Chess

Chess [77] is a reconfigurable system developed by HP labs. The reconfigurable unit is called RRA (reconfigurable arithmetic array), and it is intended to be used with multimedia applications.

3.2.13.1 RU Coupling, Reconfigurable System, Granularity, Instruction Type, Reconfiguration and Execution

The RRA is composed of 4-bit ALUs. Each ALU can perform 16 different functions (such as addition, subtraction and logical ones), generating a result of 4 bits plus a carry output. It is possible to cascade them in order to achieve wider word

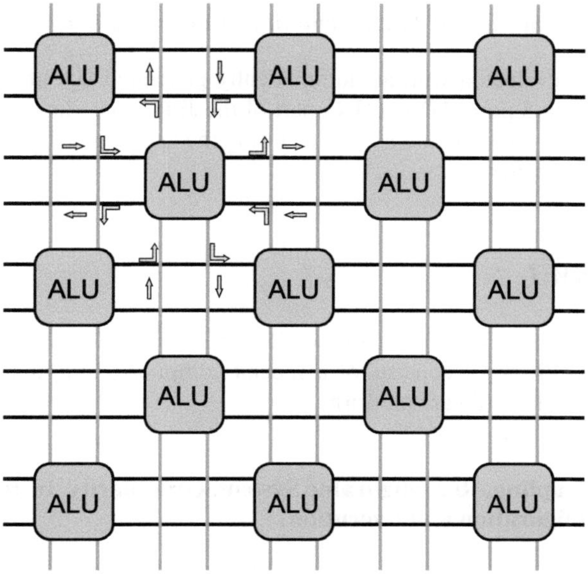

Fig. 3.23 How the ALUs are organized in Chess

sizes. These instructions can be constant or dynamic. Constant means that they are static during the whole execution, and are part of the configuration. The ALUs are connected to each other through a bus of 4 bits. The responsible components for the interconnection are the switchboxes. Each ALU is adjacent to four switchboxes, and vice-versa. This way, each ALU has input and output buses on all four sides, and is able to communicate to any of the eight surrounding ALUs, as demonstrated in Fig. 3.23.

Switchboxes can work in two different modes: a cross point switch with 64 connections (64 bits are necessary for configuring it), so it can be connected to two buses (vertical and horizontal) that pass over them; or they can be employed as a small RAM, making use of the 64 bits initially used for keeping the configuration. However, if a large number of switchboxes is used as RAM, the routing capability of the array will be reduced. Although the grain is coarser than traditional FPGAs, this architecture can still be considered as fine grain, since it works at the nibble (4 bits) level (coarse grain architectures usually work with at least 8 bits at a time).

Memory blocks are distributed throughout the basic array of ALUs and switchboxes. The baseline design provides one RAM of 256 words (8 bytes each) per 16 ALUs. Each block RAM takes approximately the same area as 4 ALUs and switchboxes. To achieve higher clock speed factors, CHESS provides two registers/buffers per switchbox (in addition to the register in each ALU), so it is possible to heavily pipeline long connections. CHESS does not support partial reconfiguration.

3.2.13.2 Code Analysis and Transformation, and Evaluation

The system was evaluated with an automatically generated layout running an implementation of an 8 point 1D IDCT (a part of the JPEG decoding algorithm). The process of building the structure as well as fitting the application in the system must be done by the designer.

3.2.14 PRISM I

PRISM [39] is an acronym for Processor Reconfiguration through Instruction Set Metamorphosis. PRISM-I consists of a special compiler (configuration compiler) and a reconfigurable hardware platform.

3.2.14.1 RU Coupling, Reconfigurable System, Granularity, Instruction Type, Reconfiguration and Execution

PRISM-I is loosely coupled, fine-grain (FPGA based) reconfigurable system. PRISM-I platform is composed of a processor board with a processor (Armstrong), which is a version of the M68010, and a second board with four Xilinx 3090 FPGAs. They are interconnected through the M68010 coprocessor interface.

3.2.14.2 Code Analysis and Transformation

The special compiler receives as input a high level language, producing two images: one for the hardware and another for the software. The hardware image is composed of a set of specifications to be used with the reconfigurable platform. It is generated from C constructs that are transformed to a hardware specification language, which will be mapped to a XILINX FPGA. According to the authors, it uses a similar technique when comparing to automatic synthesis to silicon.

3.2.14.3 Evaluation

Several functions were implemented, such as: hamming (24 times faster when comparing to the host processor), bit reversal, a digital logic circuit simulator (almost 7 times faster).

3.2.15 PRISM II

PRISM I was considered as a "proof of concept" by the authors. This way, PRISM II was built to run real life applications, although the principles are the same of the previous architecture [73].

3.2.15.1 RU Coupling

The system is loosely coupled to the AMD Am29050 processor.

3.2.15.2 Reconfigurable System, Granularity, Instruction Type, Reconfiguration and Execution

The main goals of PRISM II were: hardware synthesis of functions from a subset of C language; support for sequential logic in this synthesis (PRISM I just supports combinational logic) so more constructions, such as loops, can be mapped; optimizations on data transfer between the RU and GPP; in opposite to PRISM-I, be cost effective in terms of hardware. In addition, more optimizations were implemented when compared to PRISM I. For instance, PRISM II is faster concerning context switching in the RU.

The PRISM hardware platform is shown in Fig. 3.24. As can be observed, more than one FPGA (in this case, Xilinx 4010) can be used. Data buffers are responsible for the management of data exchange. The global bus is used to provide control signals as well data exchange between the FPGAs. The communication between the GPP and the reconfigurable units is done using the co-processor interface of the Am29050. Data transferred to and from FPGAs can be in different quantities: 8, 16 or 32 bits.

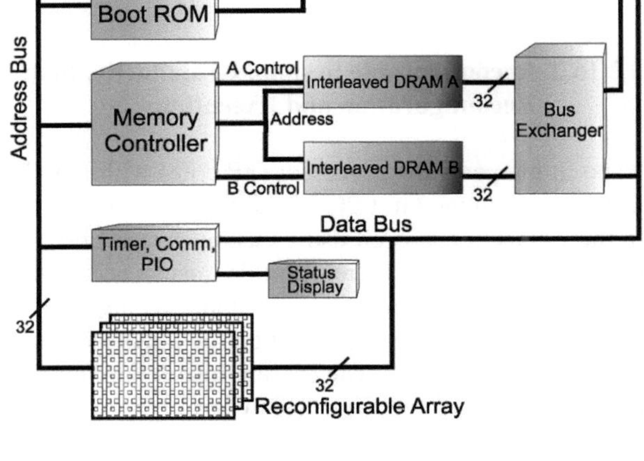

Fig. 3.24 Block Diagram of the PRISM-II Reconfigurable System

Table 3.6 Speedups using two different hardware synthesis programs

Benchmark	Speedup VHDL Designer	Speedup X-BLOX
Piece	6.67	9.22
Hamming D.	23.39	23.01
Bit Reversal	86.3	83.19
Logic F.	7.38	-
Search	46.98	46.25
Hamming C.	10.01	10.38

3.2.15.3 Code Analysis and Transformation

The configuration compiler of PRISM-II has two components. The first one is a C parser and optimizer, based on GCC. From this intermediate generated program file, both control and data flows graphs are constructed. These graphs are used for hardware synthesis (the second component).

3.2.15.4 Evaluation

The system was evaluated with the following benchmarks: Piecewise linear approximation to a function, Hamming distance and computation, Bit reversal, Mix of logic functions, and bit search. The speedups when comparing against the standalone processor are reported in Table 3.6. Two different programs for the hardware synthesis were employed: "VHDL Designer" and "X-Blox".

3.2.16 Nano

3.2.16.1 RU Coupling

The Nano system [103] consists of a reconfigurable logic tightly coupled to an accumulator-based processor, both sharing the same processor resources.

3.2.16.2 Reconfigurable System, Granularity, Instruction Type, Reconfiguration and Execution

Nano implements the processor (nP core) within an FPGA, so the processor itself is also programmed in FPGA (although it is not configurable). The reconfiguration in the FPGA is achieved using custom instructions. The nP core is a general purpose processor. It has no register file, just an accumulator. With its pipeline of three stages (Instruction Fetch, Decode and Execution), it implements just six different instructions. They have fixed length of 2 bytes each, divided in two parts: opcode and operand reference. The custom instruction modules interface with the nP core using the registers and control signals.

3.2.16.3 Code Analysis and Transformation

New custom instructions are developed with high level synthesis tools. After a custom instruction is built, it goes to a library, so it can be reused in future designs. User must program the nP in assembly, using the core nP instructions or the custom ones previously implemented. This way, the development of custom instruction as well as the decoder responsible for them should be programmed by the designer. An integrated assembler is responsible for generating the executable with instructions of both processor and custom logic.

3.2.16.4 Evaluation

The system was implemented with FPGA based on the Xilinx 3000 series and was evaluated with an nP configuration designed to control a sound card, which has three main functions: transfer of PCM audio, handle asynchronous data transfer and control the external synthesis engine. Five modules were added to the nP to handle with the following interfaces: MIDI, Codec, PC, Synthesis and Memory.

3.3 Recent Dataflow Architectures

More recently, new dataflow architectures were proposed. These architectures abandon program counter and the linear von-Neumann execution that could limit the amount of parallelism to be explored. These systems are highly dependent on compilers and tools to code generation, which involves placing parts of the code in the correct order in the processing elements, resolve synchronism, parallelism analysis and other aspects of the runtime environment. This way, the main effort is on the development of these tools, not on the hardware design, which is usually very simple and regular.

As a first example, TRIPS [58, 87] is a hybrid von-Neumann/dataflow architecture that combines an instance of coarse-grained, polymorphous grid processor cores with an adaptive on-chip memory system. To better explore the application parallelism and provide a large use of available resources, TRIPS uses three different modes of execution: D-morph, which explores parallelism in instruction level; T-morph, which works at the thread level, mapping multiple threads onto a single TRIPS core; and S-morph, which is targeted to applications like streaming media with high data-level parallelism. Figures 3.25 and 3.26 give an overview of the TRIPS architecture.

Another example of a dataflow machine is Wavescalar [90, 91] that, likewise TRIPS, relies on the compiler to statically allocate instructions into its hardware structures. As it can be observed in Fig. 3.27, the basic processing element is very similar to the one found in TRIPS. However, this architecture is even more regular when considering its structure.

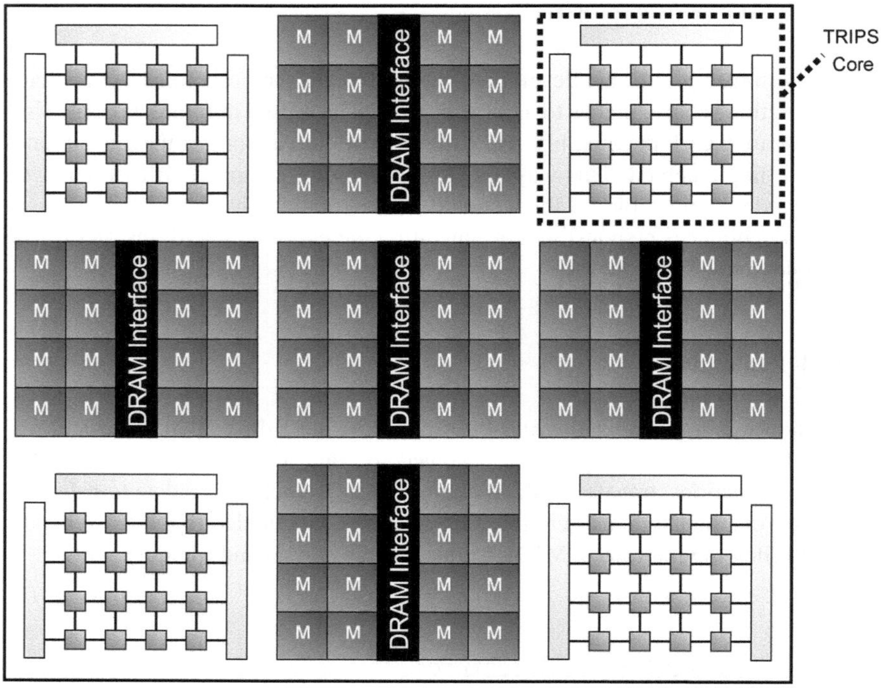

Fig. 3.25 General overview of the TRIPS architecture: the TRIPS Chip

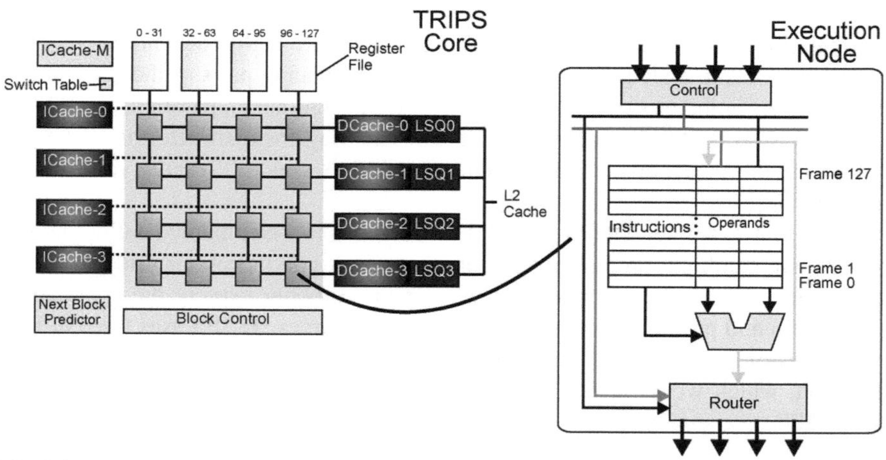

Fig. 3.26 The TRIPS core, and an execution node

In the same work, the motivations for building a dataflow architecture are discussed. They are related to some limitations that superscalar processors present, mainly because they are Von-Neumann oriented architectures. The so-called processor scaling wall is discussed, which emerges because of the following reasons:

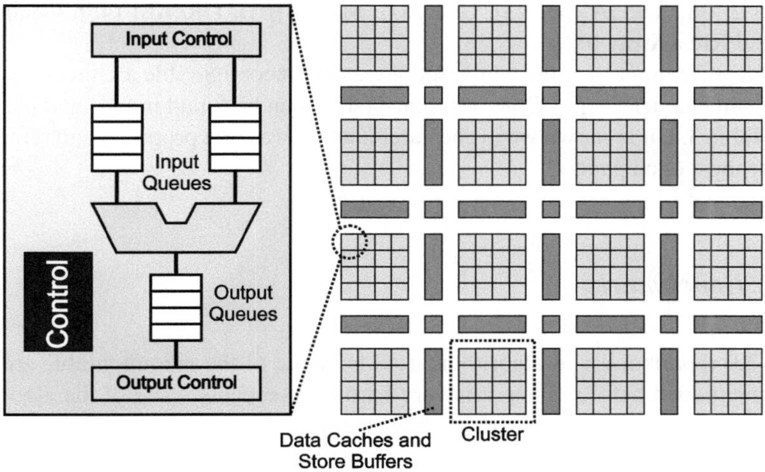

Fig. 3.27 The Wavescalar architecture

- The difference in terms of speed between (fast) transistors and (slow) wires is increasing, meaning that there is a disparity between computation and communication;
- The increasing cost of circuit complexity;
- The decreasing reliability of these circuits.

According to the authors, superscalar processors will be the main victims, since they have a huge infrastructure with slow broadcast networks, associative searches, complex control logic and inherently centralized structures. Moreover, other drawbacks regarding superscalar architectures can be cited:

- They have an inherent complexity, making efficient implementations a challenge;
- Their execution model centers around instruction fetch. It is an intrinsic serialization point.

On the other hand, dataflow machines must convert control dependences into data dependences. To accomplish this, they explicitly send data values to the instructions that need them, instead of broadcasting these data values via the register file. The potential consumers are known at compile time.

3.4 Summary and Comparative Tables

3.4.1 Other Reconfigurable Architectures

Other reconfigurable systems are worth to be briefly cited in this section, such as: DISC [102], Pleiades [106], Montium [66], XiRISC [75], ReRISC [96], Napa [85],

Splash 2 [38], DPGA [55, 93], Colt [67], Matrix [81], DReAM [45], Chameleon [86] and KressArray [62].

Table 3.7 summarizes the whole set of existent reconfigurable architectures, with their main characteristics. Other reference tables can be found in [41] and [61].

In Table 3.8, it is shown when each architecture was first proposed and references to their most cited papers.

3.4.2 Benchmarks

Table 3.9 reports a list of algorithms used by some of the reconfigurable architectures mentioned before. Since it is very hard to re-execute some of the algorithms because of the lack of information about them, and no access to the exact source code is available, one will briefly discuss the behavior of these algorithms, based on the articles written by the associated authors or on works that have already used somehow the same benchmarks.

REMARC executes MPEG2 decoding, optimizing just two kernels: IDCT and MC that, according to the authors, cover more than 70% of the total execution time. It also executes MPEG2 encoding, optimizing the Motion Estimation, which covers 98% of total execution time. The third algorithm employed is DES. DES is a well-known cryptography algorithm and, as can be observed in Fig. 3.28, its structure facilitates its implementation in reconfigurable logic, for its regularity and simplicity of basic operations to be performed.

GARP, in turn, was evaluated with the following algorithms:

- *DES*: Just discussed before.
- *Sorting*: Different types of sorting, including Quicksort (1 million objects). Sorting algorithms usually have just one kernel used for sorting the components, which is repeated several times.
- *Image Dithering*: It is a vector processing algorithm. Again, it is based on the same kernel that is repeated several times through the image. In this case, a dither was applied to a full color image of 640 × 480 pixels to a fixed palette of fewer than 256 colors.

The following algorithms were executed on the CHIAMERA system:

- *Compress/SPEC92*: It is interesting that, opposite to the previous algorithms, this one presented a speedup of just 1.11. This can be explained because it is very likely that there are no distinct kernels for optimization: in the work of [74], in a study on value prediction, the authors show that this algorithm has a small value locality concerning loads, which could be a reflex of a small reutilization of kernels.
- *Eqntott/SPEC92*: According to the paper, this program spends about 85% of its time in a single routine, *cmppt*. In the same work on value prediction cited before, Eqntott has a high degree of load value locality.

Table 3.7 General characteristics of several reconfigurable architectures

Name	GPP	Coupling	Granu-larity	Code Analysis/ Transformation
Chimaera	MIPS R400	Tightly	Fine	Hand-coded/modified GCC
GARP	MIPS II	Loosely	Fine	Hand-coded/modified GCC + special language
REMARC	MIPS	Loosely	Fine	Hand-coded/modified GCC + special language
Rapid	Standalone	-	Coarse	Modified C language
Piperench	-	Loosely	Coarse	Parameterized Compiler
Molen	PowerPC	Loosely	Fine	Hand-coded/modified GCC
Morphosys	TinyRISC	Loosely	Coarse	Hand-coded/Special Compiler
ADRES	VLIW	Tightly	Coarse	Framework (IMPACT + VLIW compiler)
Concise	MIPS-like RISC	Tightly	Fine	Automated search for hotspots/ C Translator to HDL
PACT-XPP	Standalone	-	Coarse	Special hardware language (NML)/Modified C
RAW	Standalone	-	Coarse	Modified C Compiler
Onechip	MIPS	Tightly	Fine	Modified C Compiler
PRISM II	AMD Am29050	Loosely	Fine	Hand-coded/modified GCC
Chess	Standalone	-	Coarse	Hand-coded
Nano	nP Core	Tightly	Fine	Hand-coded
DISC	-	Loosely	Fine	Hand-coded
Montium	Standalone	-	Coarse	Automated modified C to Montium Design Flow
XiRISC	RISC VLIW	Tightly	Fine	Dedicated Framework/GCC
Pleiades	ARM8	Loosely	Mixed	Hand-coded
ReRISC	RISC	Tightly	Coarse	Automated Search/Transformation
Napa	Compact RISC	Loosely	Fine	Dedicated Framework/special language based on C
Splash	Standalone	-	Fine	Hand-coded
DPGA	Standalone	-	Fine	Hand-coded
Colt	Standalone	-	Coarse	Hand-coded
Matrix	Standalone	-	Coarse	Hand-coded
DreAM	Standalone	-	Coarse	Hand-coded
Chamaleon	ARC Processor	Loosely	Coarse	Specific Framework/C-Based design

Table 3.8 First year of publication and references

Name	Year of First Pub.	References	Name	Year of First Pub.	References
PRISM	1993	[39]	Napa	1998	[85]
PRISM II	1993	[73]	Piperench	1999	[51] [54] [60]
Splash 2	1993	[38]	Concise	1999	[68]
Nano	1994	[103]	Chess	1999	[77]
DPGA	1995	[55][93]	DreAM	2000	[45]
DISC	1995	[102]	KressArray	2000	[62]
Colt	1996	[67]	Pleiades	2000	[106]
Onechip	1996	[49] [103]	Chamaleon	2001	[86]
Rapid	1996	[52] [53] [56]	Molen	2001	[98] [99]
Matrix	1996	[81]	ADRES	2002	[79]
Chimaera	1997	[63] [64]	PACT-XPP	2003	[44]
GARP	1997	[46] [65]	Motium	2003	[66]
RAW	1997	[94] [100]	XiRISC	2003	[75]
Morphosys	1998	[71] [88] [89]	Pleiades	2000	[106]
REMARC	1998	[82] [83]	ReRISC	2005	[96]

- *Conway's Game of Life*: According to the paper, it is basically an array computation. In the software version of the algorithm, more than half of the time is spent in the routines *getbit* and *putbit*, which perform the reads and writes of the value of individual cells.

RaPiD, in turn, executes only two algorithms: FIR filter and Matrix Multiplication. These algorithms are easily programmable and are highly based on just one kernel with almost no control instructions at all, proving once more that traditional reconfigurable architectures in general just attack one niche of applications.

To evaluate PipeRench's performance, the authors have also chosen dataflow oriented software, dominated by very distinct kernels, which is a characteristic of algorithms that are highly based on filters or transforms, as can be observed:

- *Automatic target recognition* (*ATR*): shape-sum kernel of the Sandia algorithm for automatic target recognition;
- *Cordic*: Honeywell timing benchmark for Cordic vector rotations;
- *DCT*: 1D, 8-point discrete cosine transform;
- *DCT-2D*: 2D discrete cosine transform;
- *FIR*: finite-impulse response filter, with 20 taps and 8-bit coefficients;
- *IDEA*: complete 8-round International Data Encryption Algorithm;
- *Nqueens*: an evaluator for the N queens problem on an 8×8 board;
- *Over*: the Porter-Duff over operator;
- *PopCount*: a custom instruction implementing a population count instruction;

Molen was evaluated with the MPEG2 encoder/decoder. The most time consuming operations among SAD (sum of absolute difference), 2D-DCT (two dimensional

Table 3.9 Benchmark executed on the most popular reconfigurable architectures

	Remarc	Garp	Chimaera	Morphosys	RaPiD	PipeRench	Molen	Adres	Onechip	TRIPS	PACT-XPP
DES enc/dec	X	X	X								
Simple Gausian Blur			X								
RGB-Grayscale conv			X								
RC5 64/8/12 enc			X								
Game of Life			X								
Eqntott			X								
Skeletonization			X								
DNA String Comp			X								
MPEG2 Encoder	X		X				X				
ADPCM encoder			X								
Compress			X								
Motion Estimation				X	X						
DCT-2D 8x8				X	X	X			X		
ATR				X		X					
IDEA				X		X				X	
Cordic						X					
DCT						X			X		
FIR					X	X		X	X	X	X
Nqueens						X					
Over						X					
PopCount						X					
Matrix Multiply					X			X			
Convert(RGB to YIQ)										X	
FFT										X	
Transform										X	
IQ e IDCT 8x8											X
Sort		X									
Image Dithering		X									
Image median filter		X									
Strlen		X									
MPEG2 Decoder	X						X	X			
JPEG							X				
IDCT								X			
ADPCM Decoder								X			

discrete cosine transform), and 2D-IDCT (two dimensional inverse DCT) were optimized. These kernels, in turn, are the most time consuming ones in the MPEG2 algorithm and highly dataflow oriented [69].

ADRES, OneChip, TRIPS and XPP basically execute the very same class of algorithms, as can be observed in the same table: they are encoders or decoders, matrix operations or filters. This way, it is clear that the design of a reconfigurable system able to optimize any kind of algorithm is a task to be handled.

Reinforcing this idea, it is very interesting to note that almost the totally of the referenced works about reconfigurable architectures, analyzed in Sect. 3.1, employ as a benchmark set exactly the ones which have very distinct kernels subject of op-

Fig. 3.28 Steps of the DES algorithm [83]

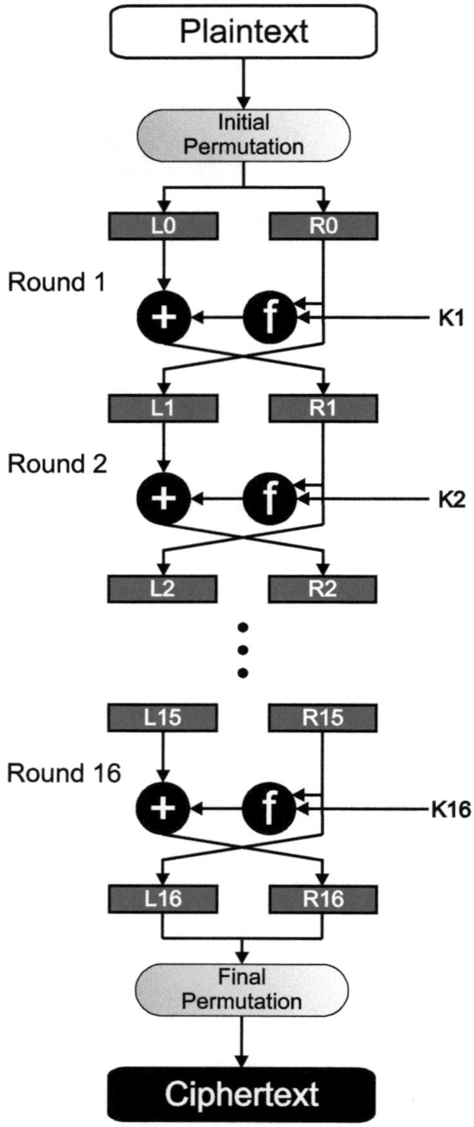

timization, and those that are very dataflow oriented. As previously discussed, these two characteristics make these benchmarks the most suitable ones for execution in reconfigurable fabric. They correspond to just one area in a graph considering two axis (number of distinct kernels and control/dataflow behavior), as illustrated in Fig. 3.29. As explored in Sect. 2.6.1, this situation is far from being the reality of embedded systems and, of course, of the general purpose computation field.

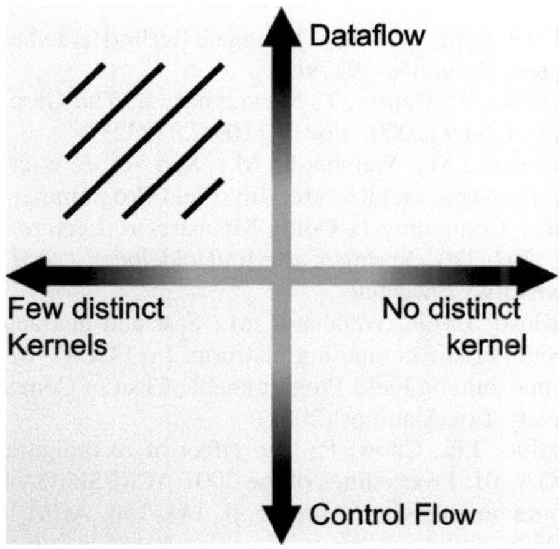

Fig. 3.29 Different groups of behaviors: reconfigurable systems usually attack just one niche

This way, reconfigurable systems are only efficient for a determined field of application. To make this scenario even worse, the new era of embedded systems gives to the user the opportunity of installing and executing different applications, whose behavior is non-predictable during the production of the devices, as in the general purpose computation. This lack of flexibility can be solved only through the use of dynamic optimization: the system's ability to adapt itself during execution. This will be the subject of the next chapter.

References

38. Arnold, J.M., et al.: The splash 2 processor and applications. In: International Conference on Computer Design. CS Press, München (1993)
39. Athanas, P.M., Silverman, H.F.: Processor reconfiguration through instruction-set metamorphosis. Computer **26**(3), 11–18 (1993). doi:10.1109/2.204677
40. Babb, J., Frank, M., Lee, V., Waingold, E., Barua, R., Taylor, M., Kim, J., Devabhaktuni, S., Agarwal, A.: The raw benchmark suite: computation structures for general purpose computing. In: FCCM '97: Proceedings of the 5th IEEE Symposium on FPGA-Based Custom Computing Machines, p. 134. IEEE Computer Society, Los Alamitos (1997)
41. Barat, F., Lauwereins, R.: Reconfigurable instruction set processors: A survey. In: RSP '00: Proceedings of the 11th IEEE International Workshop on Rapid System Prototyping (RSP 2000), p. 168. IEEE Computer Society, Los Alamitos (2000)
42. Barua, R., Lee, W., Amarasinghe, S., Agarwal, A.: Maps: A compiler-managed memory system for raw machines. In: Proceedings of the 26th International Symposium on Computer Architecture, pp. 4–15 (1998)
43. Barua, R., Lee, W., Amarasinghe, S., Agarwal, A.: Memory bank disambiguation using modulo unrolling for raw machines. In: Proceedings of the ACM/IEEE Fifth Int'l Conference on High Performance Computing (HIPC) (1998)
44. Baumgarte, V., Ehlers, G., May, F., Nückel, A., Vorbach, M., Weinhardt, M.: Pact xpp—a self-reconfigurable data processing architecture. J. Supercomput. **26**(2), 167–184 (2003). doi:10.1023/A:1024499601571
45. Becker, J., Pionteck, T., Glesner, M.: DReAM: ADynamicallyReconfigurable architecture for future mobile communication applications. In: Field-Programmable Logic and Applications: The Roadmap to Reconfigurable Computing. Lecture Notes in Computer Science,

vol. 1896, pp. 312–321. Springer, Berlin/Heidelberg (2000). http://www.springerlink.com/content/3kvjdm6qxf9k7xt5/
46. Callahan, T., Hauser, J., Wawrzynek, J.: The Garp architecture and C compiler. Computer **33**(4), 62–69 (2000). doi:10.1109/2.839323
47. Cardoso, J.M., Weinhardt, M.: Xpp-vc: A c compiler with temporal partitioning for the pact-xpp architecture. In: Field-Programmable Logic and Applications: Reconfigurable Computing Is Going Mainstream. Lecture Notes in Computer Science, vol. 2438, pp. 207–226. Springer, Berlin/Heidelberg (2002). http://www.springerlink.com/content/2rwnvdfwv79wev9u/
48. Cardoso, J.M.P., Weinhardt, M.: Fast and guaranteed c compilation onto the PACT-XPP reconfigurable computing platform. In: FCCM '02: Proceedings of the 10th Annual IEEE Symposium on Field-Programmable Custom Computing Machines, p. 291. IEEE Computer Society, Los Alamitos (2002)
49. Carrillo, J.E., Chow, P.: The effect of reconfigurable units in superscalar processors. In: FPGA '01: Proceedings of the 2001 ACM/SIGDA Ninth International Symposium on Field Programmable Gate Arrays, pp. 141–150. ACM, New York (2001). doi:10.1145/360276.360328
50. Chang, P.P., Mahlke, S.A., Chen, W.Y., Warter, N.J., Hwu, W.M.W.: Impact: an architectural framework for multiple-instruction-issue processors. SIGARCH Comput. Archit. News **19**(3), 266–275 (1991). doi:10.1145/115953.115979
51. Chou, Y., Pillai, P., Schmit, H., Shen, J.P.: Piperench implementation of the instruction path coprocessor. In: MICRO 33: Proceedings of the 33rd Annual ACM/IEEE International Symposium on Microarchitecture, pp. 147–158. ACM, New York (2000). doi:10.1145/360128.360144
52. Cronquist, D.C., Fisher, C., Figueroa, M., Franklin, P., Ebeling, C.: Architecture design of reconfigurable pipelined datapaths. In: ARVLSI '99: Proceedings of the 20th Anniversary Conference on Advanced Research in VLSI, p. 23. IEEE Computer Society, Los Alamitos (1999)
53. Cronquist, D.C., Franklin, P., Berg, S.G., Ebeling, C.: Specifying and compiling applications for rapid. In: FCCM '98: Proceedings of the IEEE Symposium on FPGAs for Custom Computing Machines, p. 116. IEEE Computer Society, Los Alamitos (1998)
54. David, H.S., Whelihan, D., Tsai, A., Moe, M., Levine, B., Taylor, R.R.: Piperench: A virtualized programmable datapath in 0.18 micron technology. In: Proc. of IEEE Custom Integrated Circuits Conference, pp. 63–66 (2002)
55. DeHon, A.: Dpga utilization and application. In: FPGA '96: Proceedings of the 1996 ACM Fourth International Symposium on Field-Programmable Gate Arrays, pp. 115–121. ACM, New York (1996). doi:10.1145/228370.228387
56. Ebeling, C., Cronquist, D.C., Franklin, P.: Rapid—reconfigurable pipelined datapath. In: FPL '96: Proceedings of the 6th International Workshop on Field-Programmable Logic, Smart Applications, New Paradigms and Compilers, pp. 126–135. Springer, London (1996)
57. Ebeling, C., Fisher, C., Xing, G., Shen, M., Liu, H.: Implementing an ofdm receiver on the rapid reconfigurable architecture. IEEE Trans. Comput. **53**(11), 1436–1448 (2004). doi:10.1109/TC.2004.98
58. Gebhart, M., Maher, B.A., Coons, K.E., Diamond, J., Gratz, P., Marino, M., Ranganathan, N., Robatmili, B., Smith, A., Burrill, J., Keckler, S.W., Burger, D., McKinley, K.S.: An evaluation of the trips computer system. In: ASPLOS '09: Proceeding of the 14th International Conference on Architectural Support for Programming Languages and Operating Systems, pp. 1–12. ACM, New York (2009). doi:10.1145/1508244.1508246
59. Goldstein, S.C., Schmit, H., Budiu, M., Cadambi, S., Moe, M., Taylor, R.R.: Piperench: A reconfigurable architecture and compiler. Computer **33**(4), 70–77 (2000). doi:10.1109/2.839324
60. Goldstein, S.C., Schmit, H., Moe, M., Budiu, M., Cadambi, S., Taylor, R.R., Laufer, R.: Piperench: a co/processor for streaming multimedia acceleration. In: ISCA '99: Proceedings of the 26th Annual International Symposium on Computer Architecture, pp. 28–39. IEEE Computer Society, Los Alamitos (1999). doi:10.1145/300979.300982

61. Hartenstein, R.: Coarse grain reconfigurable architecture (embedded tutorial). In: ASP-DAC '01: Proceedings of the 2001 Conference on Asia South Pacific Design Automation, pp. 564–570. ACM, New York (2001). doi:10.1145/370155.370535
62. Hartenstein, R., Herz, M., Hoffmann, T., Nageldinger, U.: KressArray xplorer: a new cad environment to optimize reconfigurable datapath array. In: ASP-DAC '00: Proceedings of the 2000 Asia and South Pacific Design Automation Conference, pp. 163–168. ACM, New York (2000). doi:10.1145/368434.368597
63. Hauck, S., Fry, T.W., Hosler, M.M., Kao, J.P.: The chimaera reconfigurable functional unit. In: FCCM '97: Proceedings of the 5th IEEE Symposium on FPGA-Based Custom Computing Machines, p. 87. IEEE Computer Society, Los Alamitos (1997)
64. Hauck, S., Fry, T.W., Hosler, M.M., Kao, J.P.: The chimaera reconfigurable functional unit. IEEE Trans. Very Large Scale Integr. Syst. **12**(2), 206–217 (2004). doi:10.1109/TVLSI.2003.821545
65. Hauser, J.R., Wawrzynek, J.: Garp: a mips processor with a reconfigurable coprocessor. In: FCCM '97: Proceedings of the 5th IEEE Symposium on FPGA-Based Custom Computing Machines, p. 12. IEEE Computer Society, Los Alamitos (1997)
66. Heysters, P., Smit, G., Molenkamp, E.: A flexible and energy-efficient coarse-grained reconfigurable architecture for mobile systems. J. Supercomput. **26**(3), 283–308 (2003). doi:10.1023/A:1025699015398
67. Jr, R.B., Athanas, P.M., Musgrove, M.D.: Colt: An experiment in wormhole run-time reconfiguration. In: High-Speed Computing, Digital Signal Processing, and Filtering Using Reconfigurable Logic, pp. 187–194 (1996)
68. Kastrup, B., Bink, A., Hoogerbrugge, J.: Concise: A compiler-driven cpld-based instruction set accelerator. In: FCCM '99: Proceedings of the Seventh Annual IEEE Symposium on Field-Programmable Custom Computing Machines, p. 92. IEEE Computer Society, Los Alamitos (1999)
69. Kuzmanov, G., Gaydadjiev, G., Vassiliadis, S.: The Molen processor prototype. In: FCCM '04: Proceedings of the 12th Annual IEEE Symposium on Field-Programmable Custom Computing Machines, pp. 296–299. IEEE Computer Society, Los Alamitos (2004)
70. Lee, C., Potkonjak, M., Mangione-Smith, W.H.: MediaBench: a tool for evaluating and synthesizing multimedia and communications systems. In: MICRO 30: Proceedings of the 30th Annual ACM/IEEE International Symposium on Microarchitecture, pp. 330–335. IEEE Computer Society, Los Alamitos (1997)
71. Lee, M.H., Singh, H., Lu, G., Bagherzadeh, N., Kurdahi, F.J., Filho, E.M.C., Alves, V.C.: Design and implementation of the morphosys reconfigurable computing processor. J. VLSI Signal Process. Syst. **24**(2/3), 147–164 (2000). doi:10.1023/A:1008189221436
72. Lee, W., Barua, R., Frank, M., Srikrishna, D., Babb, J., Sarkar, V., Amarasinghe, S.: Space-time scheduling of instruction-level parallelism on a raw machine. SIGOPS Oper. Syst. Rev. **32**(5), 46–57 (1998). doi:10.1145/384265.291018
73. Lee, W.A., Agarwal, L., Lee, T., Smith, A., Lam, E., Athanas, P., Ghosh, S.: Prism-ii compiler and architecture (1993)
74. Lipasti, M.H., Wilkerson, C.B., Shen, J.P.: Value locality and load value prediction. In: ASPLOS-VII: Proceedings of the Seventh International Conference on Architectural Support for Programming Languages and Operating Systems, pp. 138–147. ACM, New York (1996). doi:10.1145/237090.237173
75. Lodi, A., Toma, M., Campi, F., Cappelli, A., Guerrieri, R.: A vliw processor with reconfigurable instruction set for embedded applications. IEEE J. Solid State Circuits **38**(11), 1876–1886 (2003)
76. Maheswaran, K., Akella, V.: Hazard-free implementation of the self-timed cell set in a xilinx fpga. Tech. Rep., University of California (1994)
77. Marshall, A., Stansfield, T., Kostarnov, I., Vuillemin, J., Hutchings, B.: A reconfigurable arithmetic array for multimedia applications. In: FPGA '99: Proceedings of the 1999 ACM/SIGDA Seventh International Symposium on Field Programmable Gate Arrays, pp. 135–143. ACM, New York (1999). doi:10.1145/296399.296444

78. Mei, B., Veredas, F.J., Masschelein, B.: Mapping an h.264/avc decoder onto the address reconfigurable architecture. In: International Conference on Field Programmable Logic and Applications, pp. 622–625 (2005). doi:10.1109/FPL.2005.1515799
79. Mei, B., Vernalde, S., Verkest, D., De Man, H., Lauwerein, R.: ADRES: An architecture with tightly coupled VLIW processor and coarse-grained reconfigurable matrix. In: Field-Programmable Logic and Applications. Lecture Notes in Computer Science, vol. 2778, pp. 61–70. Springer, Berlin/Heidelberg (2003). doi:10.1007/b12007. http://www.springerlink.com/content/03yt3xeh60r8971k/
80. Mei, B., Vernalde, S., Verkest, D., De Man, H., Lauwereins, R., Mei, B., Vernalde, S., Verkest, D., De, H., Lauwerein, R.: Dresc: A retargetable compiler for coarse-grained reconfigurable architectures (2002)
81. Mirsky, E., DeHon, A.: Matrix: A reconfigurable computing architecture with configurable instruction distribution and deployable resources. In: IEEE Symposium on FPGAs for Custom Computing Machines, pp. 157–166 (1996)
82. Miyamori, T., Olukotun, K.: Remarc (abstract): reconfigurable multimedia array coprocessor. In: FPGA '98: Proceedings of the 1998 ACM/SIGDA Sixth International Symposium on Field Programmable Gate Arrays, p. 261. ACM, New York (1998). doi:10.1145/275107.275164
83. Miyamori, T., Olukotun, K.: Remarc: Reconfigurable multimedia array coprocessor. In: IEICE Transactions on Information and Systems E82-D, pp. 389–397 (1998)
84. Panainte, E.M., Bertels, K., Vassiliadis, S.: The Molen compiler for reconfigurable processors. ACM Trans. Embed. Comput. Syst. **6**(1), 6 (2007). doi:10.1145/1210268.1210274
85. Rupp, C.R., Landguth, M., Garverick, T., Gomersall, E., Holt, H., Arnold, J.M., Gokhale, M.: The napa adaptive processing architecture. In: FCCM '98: Proceedings of the IEEE Symposium on FPGAs for Custom Computing Machines, p. 28. IEEE Computer Society, Los Alamitos (1998)
86. Salefski, B., Caglar, L.: Re-configurable computing in wireless. In: DAC '01: Proceedings of the 38th Annual Design Automation Conference, pp. 178–183. ACM, New York (2001). doi:10.1145/378239.378459
87. Sankaralingam, K., Nagarajan, R., Liu, H., Kim, C., Huh, J., Ranganathan, N., Burger, D., Keckler, S.W., McDonald, R.G., Moore, C.R.: Trips: A polymorphous architecture for exploiting ilp, tlp, and dlp. ACM Trans. Archit. Code Optim. **1**(1), 62–93 (2004). doi:10.1145/980152.980156
88. Singh, H., Lee, M.H., Lu, G., Kurdahi, F.J., Bagherzadeh, N.: Morphosys: A reconfigurable architecture for multimedia applications. In: Workshop on Reconfigurable Computing at PACT, pp. 134–139 (1998)
89. Singh, H., Lee, M.H., Lu, G., Bagherzadeh, N., Kurdahi, F.J., Filho, E.M.C.: Morphosys: An integrated reconfigurable system for data-parallel and computation-intensive applications. IEEE Trans. Comput. **49**(5), 465–481 (2000). doi:10.1109/12.859540
90. Swanson, S., Michelson, K., Schwerin, A., Oskin, M.: Wavescalar. In: MICRO 36: Proceedings of the 36th Annual IEEE/ACM International Symposium on Microarchitecture, p. 291. IEEE Computer Society, Los Alamitos (2003)
91. Swanson, S., Schwerin, A., Mercaldi, M., Petersen, A., Putnam, A., Michelson, K., Oskin, M., Eggers, S.J.: The wavescalar architecture. ACM Trans. Comput. Syst. **25**(2), 4 (2007). doi:10.1145/1233307.1233308
92. Tatas, K., Siozios, K., Soudris, D.: A survey of existing fine-grain reconfigurable architectures and CAD tools. In: Fine- and Coarse-Grain Reconfigurable Computing, pp. 3–87. Springer, Dordrecht (2007). http://www.springerlink.com/content/m561311j78506281/
93. Tau, E., Chen, D., Eslick, I., Brow, J.: A first generation dpga implementation. In: Proceedings of the Third Canadian Workshop on Field-Programmable Devices, pp. 138–143 (1995)
94. Taylor, M.B., Kim, J., Miller, J., Wentzlaff, D., Ghodrat, F., Greenwald, B., Hoffman, H., Johnson, P., Lee, J.W., Lee, W., Ma, A., Saraf, A., Seneski, M., Shnidman, N., Strumpen, V., Frank, M., Amarasinghe, S., Agarwal, A.: The raw microprocessor: A computational fabric for software circuits and general-purpose programs. IEEE Micro **22**(2), 25–35 (2002). doi:10.1109/MM.2002.997877

95. Theodoridis, G., Soudris, D., Vassiliadis, S.: A survey of coarse-grain reconfigurable architectures and cad tools. In: Fine- and Coarse-Grain Reconfigurable Computing, pp. 89–149. Springer, Dordrecht (2007). http://www.springerlink.com/content/j118u3m6m225q264/
96. Vassiliadis, N., Kavvadias, N., Theodoridis, G., Nikolaidis, S.: A risc architecture extended by an efficient tightly coupled reconfigurable unit. In: International Workshop on Applied Reconfigurable Computing (ARC), pp. 41–49. Springer, Berlin (2005)
97. Vassiliadis, S., Gaydadjiev, G., Bertels, K., Panainte, E.M.: The Molen programming paradigm. In: Proceedings of the Third International Workshop on Systems, Architectures, Modeling, and Simulation, pp. 1–10 (2003)
98. Vassiliadis, S., Wong, S., Cotofana, S.: The Molen rho-mu-coded processor. In: FPL '01: Proceedings of the 11th International Conference on Field-Programmable Logic and Applications, pp. 275–285. Springer, London (2001)
99. Vassiliadis, S., Wong, S., Gaydadjiev, G., Bertels, K., Kuzmanov, G., Panainte, E.M.: The Molen polymorphic processor. IEEE Trans. Comput. **53**(11), 1363–1375 (2004). doi:10.1109/TC.2004.104
100. Waingold, E., Taylor, M., Srikrishna, D., Sarkar, V., Lee, W., Lee, V., Kim, J., Frank, M., Finch, P., Barua, R., Babb, J., Amarasinghe, S., Agarwal, A.: Baring it all to software: Raw machines. Computer **30**(9), 86–93 (1997). doi:10.1109/2.612254
101. Wilson, R.P., French, R.S., Wilson, C.S., Amarasinghe, S.P., Anderson, J.M., Tjiang, S.W.K., Liao, S.W., Tseng, C.W., Hall, M.W., Lam, M.S., Hennessy, J.L.: Suif: an infrastructure for research on parallelizing and optimizing compilers. SIGPLAN Not. **29**(12), 31–37 (1994). doi:10.1145/193209.193217
102. Wirthlin, M.J.: A dynamic instruction set computer. In: FCCM '95: Proceedings of the IEEE Symposium on FPGA's for Custom Computing Machines, p. 99. IEEE Computer Society, Los Alamitos (1995)
103. Wirthlin, M.J., Hutchings, B.L., Gilson, K.L.: The nano processor: A low resource reconfigurable processor. In: Buell, D.A., Pocek, K.L. (eds.) IEEE Workshop on FPGAs for Custom Computing Machines, pp. 23–30. IEEE Computer Society, Los Alamitos (1994). citeseer.ist.psu.edu/wirthlin94nano.html
104. Wittig, R.D., Chow, P.: Onechip: An fpga processor with reconfigurable logic. In: IEEE Symposium on FPGAs for Custom Computing Machines, pp. 126–135 (1995)
105. Ye, Z.A., Moshovos, A., Hauck, S., Banerjee, P.: Chimaera: a high-performance architecture with a tightly-coupled reconfigurable functional unit. SIGARCH Comput. Archit. News **28**(2), 225–235 (2000). doi:10.1145/342001.339687
106. Zhang, H., Prabhu, V., George, V., Wan, M., Benes, M., Abnous, A., Rabaey, J.: A 1-v heterogenous reconfigurable dsp ic for wireless baseband digital signal processing. IEEE J. Solid State Circuits **35**(11), 1697–1704 (2000)

Chapter 4
Dynamic Optimization Techniques

Abstract According to the discussion made at the end of the previous chapter, reconfigurable systems alone cannot deal with the high heterogeneous behavior of recent applications. Hence, the only solution to cope with that is to use dynamic optimization techniques, such as Binary Translation and reuse. The section about Binary translation starts with an explanation on how it works. The main concepts are clarified, as well as the main challenges that a binary translator mechanism must handle to work properly. The section ends with a detailed view of some examples of Binary Translation machines. The study on Reuse, in turn, covers diverse types: instruction reuse, value prediction and the difference between them; basic block, trace reuse and dynamic trace memoization.

4.1 Introduction

As it has been mentioned in the previous chapters, it is only by adding information available at run time that reconfigurable logic will be able to optimize different kernels of a given benchmark set. Any static solution will be limited by the tool set or by the amount of extra hardware design required. Therefore, in this chapter, two different techniques regarding dynamic optimization are analyzed: Trace Reuse and Binary Translation.

4.2 Binary Translation

4.2.1 Main Motivations

As can be observed nowadays, the top seller processors belong to a family, meaning that they are firmly connected to legacy ISAs. Some of them are more than 30 years old (as is the case of the X86 family). Although this fact brings limitations in terms of flexibility when designing a new architecture (meaning that potential increases

A.C. Schneider Beck Fl., L. Carro, *Dynamic Reconfigurable Architectures and Transparent Optimization Techniques*,
DOI 10.1007/978-90-481-3913-2_4,

in performance are reduced), the greatest advantage is the possibility of reusing the whole existing software base, mainly when this base owns a huge market share. Moreover, porting the code sometimes can be very difficult. This way, the development of new architectures that do not implement an existent ISA is very hard. That is one of the main reasons why one cannot find a large number of competitors on the processor market.

Let us consider the hypothesis of a new processor architecture with a totally different ISA appearing on the market. Existent applications would have to be rebuilt, so they could be executed on these new machines, or to take advantage of their maximum processing capabilities. If the application is simple and programmed in some standard language, the rebuilding process is usually also simple. For that, a native compiler should be used. However, for complex applications, with large code sizes and that need different tools to join all the small pieces, the rebuilding process may be near to impossible. It is necessary to remember that operating systems and their inherent complexity must be included in this set of existent software.

Several techniques have already been proposed to somehow solve this issue. The first one is called Software Interpretation. The Software Interpreter reads each instruction, one at a time, from any given application that was previously written for another architecture with a different ISA. Besides interpreting them (replicating their behavior), the interpreters also have the role of maintaining the state behavior as if the instructions were being executed on the reference architecture. However, since they add an extra layer to perform the interpretation, they are not fast, or not as fast as they could be. As main advantages, interpreters can handle some challenges that will be studied later, such as self-modifying programs, programs with no clear barrier between instruction and data code etc.

Another alternative is the use of Microcode [121]. The idea is very similar to the one used with the software interpreter, however, it is done in hardware. There is a mechanism responsible for decoding source instructions to the new architecture, making use of a microcode ROM and a state machine. It is claimed in [108] that such technique could not be used in RISC machines, since those machines would not have a microcoded hardware layer. However, Intel has been using microcode to execute the complex X86 instructions in a RISC core for a long time [129].

Finally, one can find the main topic of this section: Binary Translators (BT). The main objective of BT is the same as the use of Software Interpretation or Microcode: to keep the porting process transparent to the user, so binary code previously compiled for another source architecture with a different ISA can be executed on another one, supporting a different instruction set, with minimal effort. This way, one can implement a different architecture executing the original ISA, breaking the strong relationship between already written code and its ISA. The BT can be static or dynamic, and can be implemented in software or hardware. This will be further explained in details in the next section.

As can be observed in Fig. 4.1, according to [131], translated programs are faster than interpreters or the use of microcode; however, binary translation is slower than native compilers.

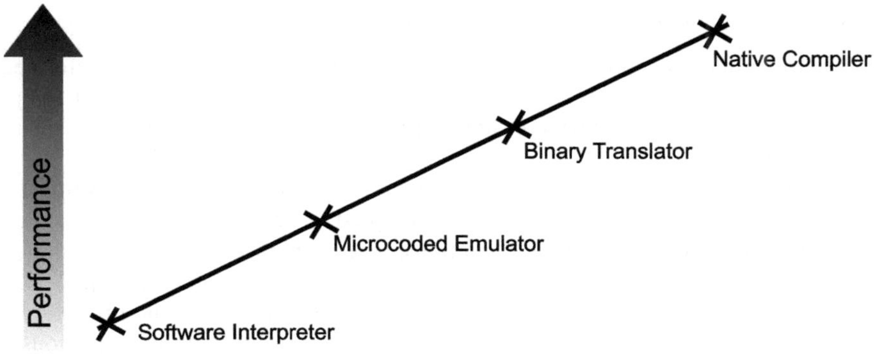

Fig. 4.1 Performance comparison between different approaches

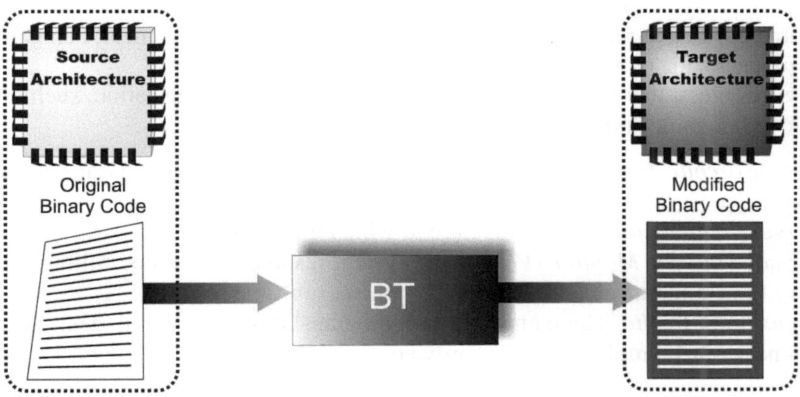

Fig. 4.2 Binary Translation Process

4.2.2 Basic Concepts

The concept of binary translation [107, 108] is very ample and can be applied in different levels. Basically, the Binary Translator is a system, which can be implemented in hardware or software, responsible for analyzing the binary code of an already compiled program. Then, some kind of transformation is done in the code, with the purpose of keeping the software compatibility (the reuse of legacy code without the need of recompilation, as illustrated in Fig. 4.2), to provide means to enhance the performance, or even both. A translated binary program is a sequence of instructions that belongs to a new architecture, and was previously translated from another ISA. According to [108], in opposite to modern high level languages, the semantics of binary code is usually well defined, so the translation is easier when applied at that level. On the other hand, without the availability of the original source code, a binary translation cannot perform some of the optimizations that a regular compiler could.

Still according to [108], there are three different kinds of binary translation:

- *Emulator*: interprets program instructions at run time. However, the transformed instructions are not saved or cached for future reuse. As it can be observed, in opposite to [131], these authors consider Software Interpretation as a type of Binary Translation.
- *Dynamic Translator*: besides interpreting the program, it saves previous translations to be used a next time, so that the overhead of translation and optimization can be amortized over multiple executions. One example of dynamic translation is just-in-time (JIT) compilers, as the ones used for Java execution [135].
- *Static Translator*: it does the job offline. Consequently, it has the opportunity of more rigorous optimization. They can also use execution profiles previously generated to achieve better results.

While in the emulator and dynamic translator approaches there are run-time overheads, since the analysis is done during program execution, static translators usually are based on a stand-alone tool that requires end-user involvement, being not as transparent as the dynamic techniques.

Nevertheless, there are other concepts regarding Binary Translation, such as the employed nomenclature [108]:

- *Source architecture*: The original (legacy) architecture *from* which translation occurs;
- *Target Architecture*: The architecture *to* which translation occurs;
- *Virtual Machine Monitor (VMM)*: the system responsible for controlling the binary translation mechanism when it is done at run-time;
- *Translation Cache*: The memory where the translations are stored. This cache is not necessarily implemented in hardware.

Another concept intrinsically connected to binary translation is dynamic optimization. While dynamic binary translation is JIT compilation of the binary code from one architecture to another, dynamic optimization concerns the run-time improvement of the code. Usually, the general term Binary Translation is also applied when both techniques are used together.

In [109] it is claimed that the use of object-oriented languages and other techniques to facilitate software development limits the optimization possibilities of the code, when considering static compiler analysis. Moreover, pre-compiled dynamic libraries are very common (such as DLLs), so applying optimizations in the whole software sometimes is just impossible. Nevertheless, there are limitations even when considering the same family of processors having the same basic ISA. For instance, let us consider a software written to be executed on the newest X86 ISA processor. The MMX or SSE multimedia instructions could not be used if one wants to execute the software on the whole X86 family of processors. Earlier processor versions that do not have these instructions implemented in their ISA could not execute the application. One alternative would be using dynamic libraries or making available different executable files. However, different versions of the same program must be built. Therefore, still according to the same authors [109], the importance of dynamic optimization through BT increases as compilers will be extremely complex and bring just modest gains in general purpose applications. Otherwise, they have to be tailored for very specific classes of applications to bring meaningful gains.

Binary translation can also produce other effects in the future, following the tendency of *write once, run everywhere*. For example, it is possible to use Binary Translation in order to perform transformations from different ISAs to a unique target architecture, so all efforts for optimization could be targeted to just that hardware.

4.2.3 Challenges

The authors in [108] make an interesting claim: since all machines are based on the Turing Model, any machine can be emulated on each other. As already discussed, besides achieving some kind of binary portability, another objective of BT is to run the translated code as efficiently as if it was running on the machine it was first designed to be executed on. However, according to the same authors, there are some challenges that must be dealt to achieve this objective:

4.2.3.1 Register Mapping

One of the basic roles of a BT mechanism is to map registers from the source to the target architecture. However, there are cases where the target architecture has less registers than the source, so some of them must be kept in memory. Memory accesses are more costly, so the distribution must be well balanced. Moreover, some architectures have state registers (for example, flags from the ALU or segmentation registers). These state-registers of the source architecture must somehow also be kept in the target architecture.

4.2.3.2 Memory Mapped I/O

This issue concerns the memory spaces that are mapped to I/O devices and that can present side effects. For instance, referencing a given memory position can inject a packet in the network, or change the status of an I/O device.

4.2.3.3 Atomic Instructions

In systems that run processes concurrently, it is very common to find some instructions that must be atomically executed in respect to memory. For example, a given memory address must be "locked" during the execution of an instruction, so a second processor could not access or modify it, handling some synchronizations issues. Replicating this behavior can be complicated. Another problem is instructions that take more than one cycle to perform their functions. These instructions must be executed completely, meaning that they cannot be interrupted. This behavior must be replicated in the target architecture.

On the other hand, in many cases, a given instruction that takes just one cycle to be executed in the source architecture takes several cycles to be executed on the target processor. In this case, the target processor cannot be interrupted until the instruction is totally executed. This way, dealing with precise exceptions can also become a problem. It can become even worse if one considers that after binary translation is applied, instructions could be reordered or even have their order changed to improve performance.

4.2.3.4 Issues Related to the Code

Several issues are related to the code. The first one is when there is no clear barrier between code and data, so it is harder to realize what parts can be modified or protected. Self-modifying code is another problem. For example, a self-modifying code changed itself, and an old version of this sequence is still cached. This means that one needs to detect whether the code was changed, so it can be invalidated in the translation cache. Furthermore, the code can try to check itself (to perform a checksum, for instance). However, if the program was modified by the BT mechanism, the checksum of the new version of the code will obviously not be right, since that verification process was expecting the original code. One solution could be to always keep a copy of the original program in a separated memory space.

4.2.3.5 OS Emulation

How to use the OS together with a BT mechanism depends on how the BT was implemented. As it will be show in the examples, some BT mechanisms work above the OS, so the OS knows about its existence, and sometimes even controls it. In this case, the OS can execute both native and translated applications. When a legacy code is detected, the translator is launched. The BT mechanism can be also found below the OS. If that is the case, it could be possible to install the OS of the source architecture and use it together with the source application in a totally transparent process. This way, the entire legacy OS code of the source architecture (or its libraries) can be emulated using the Binary Translation mechanism.

4.2.4 Examples

Besides the JIT compiler, commented before, there are other examples of the different types of Binary Translation mechanisms. The Hewlett-Packard Dynamo [109] operates entirely at runtime in order to dynamically generate optimized native re-translations of the running program's hot spots. In fact, the BT itself is an optimizing software, previously compiled and executing on the target machine. It operates transparently (any kind of interference from the user is not necessary) monitoring program behavior in order to find hot spots to be optimized, using low-overhead techniques. Then, this modified code is executed again when necessary. Operating on HP-UX, Dynamo has a code size of less than 265 Kilobytes.

Another example of the same approach, but with a different purpose, is the Compaq's FX!32 [111, 122], aimed to allow the execution of 32-bit x86 Windows applications on Alpha computers. There are other architectures that mix hardware and software to perform BT. DAISY [114, 115], from IBM, is one of those. It uses the PowerPC as source architecture and a special architecture based on a VLIW, named DAISY VLIW, as target. It is important to point out that, in opposite to Dynamo, which runs above the HPUX operating system, DAISY runs below its operating system. This way, it can be considered even more transparent to the final user, in the sense that one cannot identify it as a service or software running on the operating system.

The Transmeta Crusoe [113] shares several similar elements with DAISY. The significant difference is that Crusoe emulates an x86 system, while DAISY emulates a PowerPC. Both perform full system emulation including not only application code, but also operating systems and other privileged code. Furthermore, both use an underlying VLIW chip specifically designed to support BT as target architecture, aimed for high performance. There are also similarities regarding the optimization process: code is first interpreted and profiled and, if a fragment turns out to be frequently executed (in this case, more than 50 times), it is translated to native Crusoe instructions. Both DAISY and Crusoe are illustrated in Fig. 4.3.

Aside from the different source architectures emulated, Crusoe and DAISY differ in their intended use. DAISY is designed for use in servers and consequently is a big machine capable of issuing 8 to 16 instructions per cycle, with gigabytes of total memory. Given this large machine, the DAISY VMM emphasizes extraction of parallelism when translating from PowerPC code. DAISY reserves 100 MB or more for itself and its translations. Crusoe is aimed at low power and mobile applications such as laptops and palmtops. The processor issues only 2 to 4 instructions per cycle,

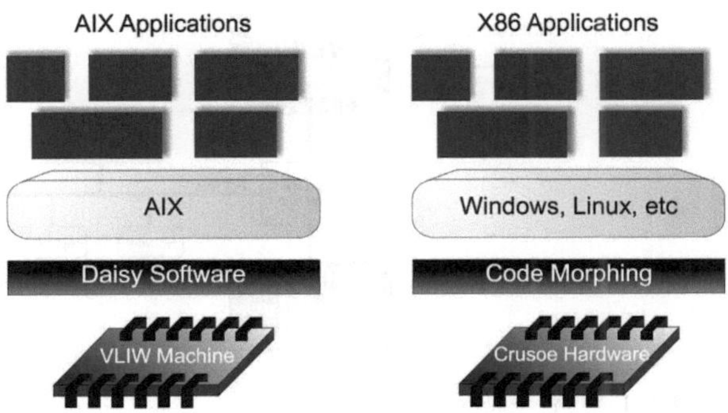

Fig. 4.3 Daisy and Transmeta Systems

and has 64 to 128 MB of total memory in a typical system. Thus, Crusoe reserves 16 MB for itself and its translation. In benchmark tests, DAISY can complete the equivalent of 3 to 4 PowerPC instructions per cycle. Transmeta has claimed that the performance of a 667-MHz Crusoe TM5400 is about the same as a 500-MHz Pentium III [130], but at a fraction of the power dissipated by the Intel machine. More details on these architectures and others will be discussed in the following sub-sections.

4.2.4.1 DAISY

DAISY (Dynamically Architected Instruction Set from Yorktown) system aims to execute PowerPC code on an eight-issue VLIW processor [114, 115, 120]. This processor is a 32-bit load-store architecture with 256-bit VLIW instructions. It has 64 registers of 32 bits each and 16 conditions registers (4 bits, each). A VLIW instruction can have up to 8 operations with three operands each, besides a control header.

Figure 4.4 shows how DAISY system is composed. The DAISY VMM code is stored in the DAISY flash ROM. When the system powers up, the VMM code is copied to the DAISY portion of memory, and the VLIW machine starts executing it. After the VMM (Virtual Machine Monitor) software initializes itself and the system, it begins translating the code of the PowerPC flash ROM to be executed on the VLIW processor. Then, this translated firmware loads the operating system (in this case, AIX Unix), which DAISY likewise translates and executes. After that, any application that is executed on the AIX can benefit from the binary translation mechanism and the VLIW processor.

This way, in the point of view of the PowerPC instructions, the BT process is totally transparent. The basic idea of the VMM is that it implements a virtual PowerPC architecture, and it is not visible to the software that is running on it. The VMM is part of the firmware, acting on the entire instruction set, including any kind of system level operation. This way, in opposite to other BT systems, DAISY is not a

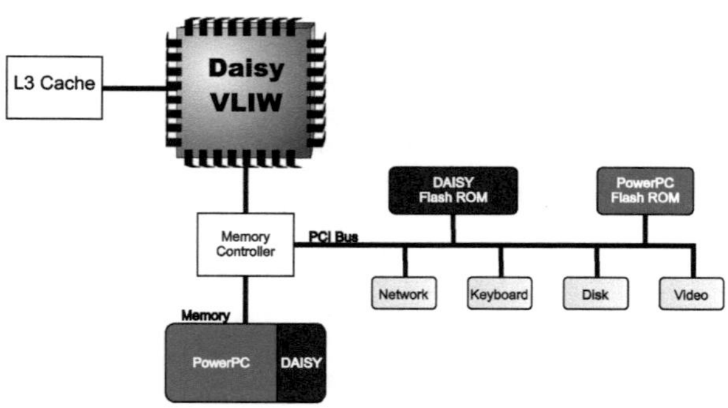

Fig. 4.4 DAISY system

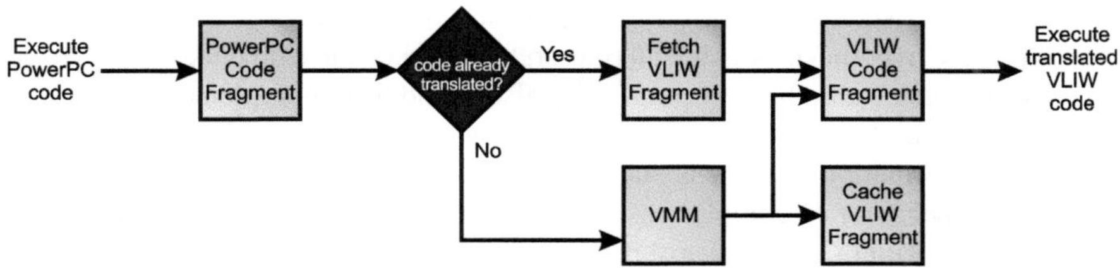

Fig. 4.5 DAISY Translation Process

part or service of the operating system running on the target machine. Consequently, DAISY can even boot an OS previously compiled for PowerPC systems.

The first time the VMM finds a code fragment of PowerPC instructions, these are interpreted. During the interpretation, the code is also profiled: this data will be used later for code generation. Considering that the purpose of the system is to achieve the maximum possible performance, the decision on whether the fragment is worth to be translated or not relies on how many times it has already been interpreted, its ILP and number of operations. After translation, the code is kept in the translation cache. The next time the same code fragment is found, it is directly executed, and no translation is required (there is an exception case, when an already translated part of the code was taken off memory because of space restrictions). The advantage of using a threshold before actual code translation is that rarely used code (such as in the initialization process) will not be translated, since the translation itself has costs associated to it.

The basic translation units are called trees. They have only one entry point, and multiple exit ones. The end of a tree generation occurs when a backward branch is found (typical of loops), or when subroutine boundaries are found. The tree regions are stored in the translation cache, a memory area only accessible to the DAISY VMM, being not visible to the system running above it. In order to facilitate the binary translation process, the DAISY processor has a special instruction called LVIA (load VLIW instruction address). It is responsible for loading VLIW instructions that correspond to a chunk of PowerPC code. If the code has already been translated and is located in memory, the instruction returns the beginning of the VLIW code. Otherwise, it calls the VMM translator, so it can translate the PowerPC code to VLIW instructions, before proper execution. These steps are demonstrated in Fig. 4.5.

Besides VLIW instruction scheduling, the DAISY BT system performs a variety of optimizations, such as ILP scheduling with data and control speculation, loop unrolling, alias analysis, load-store telescoping, dead code elimination and others, as reported in [116].

4.2.4.2 VEST

The VEST system aims to translate VAX and Ultrix MIPS images to be executed on Alpha AXP computers. Besides ensuring binary reusability, the mechanism can also show speedups when comparing to the non-translated code. The authors in [131] worked on the hypothesis of using an interpreter, but they gave up once they realized that performance would be very poor. They also found that an implementation using microcode would be inconsistent with the Alpha RISC design.

The transformation process is static and completely automatic. It is able to reproduce the behavior of complex instructions, atomicity (for the execution of multithread applications), and also arithmetic traps and error handlers. If for any reason translation is not possible, an explicit error message is given, with details on what happened.

The static translation from a VAX image involves two phases: analyzing the VAX code and the proper translation. The translated image runs with the assistance of a special environment. As the Alpha processor has more registers than the VAX, register mapping cannot be considered as a problem. Alpha has separated registers for integer and floating point operations, while in VAX there is no such distinction. This way, register mapping depends on the operation. For instance, the R1 of a VAX instruction can be mapped to the integer or floating point register sets of Alpha, depending on what operation that instruction performs. VAX condition bits are also mapped to Alpha registers.

Moreover, depending on the operation, there is tradeoff concerning performance and accuracy. For instance, it is possible to emulate the original 56-bit mantissa provided by the VAX in Alpha by software, or to use the 53-bit mantissa that is supported natively by hardware in Alpha, so it will execute faster but losing some precision. Nevertheless, there are images that cannot be translated, such as the ones that present specific cases of exception handlers, the use of undocumented system services, and software that depends on the exact memory management behavior to work properly.

As already stated, the source code can also originate from the Ultrix MIPS. This process is simpler than the VAX translation, since both source and target architectures are RISC machines. This way, many instructions can be translated on a one-to-one basis. The translation process follows two basic steps: the program is first parsed, and a graph is built; then, the code generator is called. The code generator is also responsible for register mapping and basic blocks processing. As there are not enough registers in Alpha for directing mapping, the most used source registers are kept in the target machine register file, while the others must be swapped from memory to the register bank and vice-versa. A special spill algorithm is used in order to keep the most used MIPS registers in the Alpha register bank. Some classes of code cannot be translated, such as those applications that use privileged opcodes or system calls.

Some optimizations are also performed. For example, at each subroutine call, the tool uses a pattern-matching algorithm in order to find if that subroutine corresponds to one located in a library (for instance, *strcpy*). If it is found, the call is replaced with a canned (and optimized) routine to be executed on the Alpha. Moreover, common MIPS sequences that, even though crossing basic block boundaries in MIPS, can be compressed to be executed within a single basic block in Alpha (such as the *min* and *max* functions), are also replaced with the correspondent optimized ones.

4.2.4.3 DYNAMO

The main purpose of Dynamo is dynamic optimization: it does not translate code from a source to another target architecture, but instead it optimizes a native instruction stream. The system is implemented entirely in software. Dynamo is totally transparent, since it does not depend on any kind of programmer assistance, such as user annotation or binary instrumentation. Moreover, no special compiler, OS or hardware support are required. This way, even legacy or statically optimized native binaries can be accelerated by Dynamo. A prototype was presented in [109], running on an HP PA-8000 processor, under the HPUX 10.20 OS. As Dynamo system focuses on run-time optimizations, it has the possibility of applying certain optimization techniques that would be hard for a static compiler to exploit.

Dynamo observes the running application behavior and starts the translation only when a hot spot is found. Start points for a candidate are backward taken branches, since this instruction sequence probably belongs to a loop. The exit points are branches. Each time the same candidate is found again, a counter responsible for it is incremented. When a threshold is exceeded, code generation begins. Then, Dynamo generates an optimized version of that sequence of instructions and saves it in the translation cache (called fragment cache). Dynamo only acts at that moment, since interpreting code that would not be optimized would slow down performance (Dynamo itself is a program running on the processor). When the same sequence is found again, the optimized code will be fetched from the fragment cache to be executed directly on the processor (the BT mechanism will not work at that moment). When execution finishes, Dynamo starts the whole process again. This way, the fragment cache is gradually filled with optimized application code as the execution proceeds.

Dynamo relies on the idea that a small portion of code is responsible for the majority of the application's execution time, so it is possible to benefit from the repeated reuse of optimized sequences found in the fragment cache. The prototype shows that, using Dynamo, the average performance gains considering the SpecInt95 benchmark compiled with -O is comparable to same benchmark statically compiled using -O4 (that includes extra optimizations to the specific processor), running without Dynamo.

The optimizations performed by the system are based on redundancy removal, such as branch and assignment eliminations and load removal. Other optimizations such as copy and constant propagation, strength reduction, loop invariant code motion and unrolling are also done.

Dynamo also tries to keep the translation memory occupation as low as possible, since there is the concern of maintaining the fragments in the cache and TLB. This way, there is a flushing algorithm that acts when the memory space becomes large enough to present a potential performance decrease.

Dynamo has also the concern of signal handling. Asynchronous signals are treated differently from synchronous. When the former kind arrives, it is put in a queue waiting until the fragment being executed at that moment finishes. The latter case, however, is more complex, since synchronous signals cannot be postponed.

Therefore, the implemented prototype has two kinds of code optimization: conservative and aggressive. The conservative is used when the treatment of synchronous signals is necessary. Although they not present as much performance improvements as the aggressive type, the conservative approach allows precise contexts to be constructed, so the incoming signal can be appropriately handled.

4.2.4.4 Transmeta Crusoe

Transmeta Crusoe [113] executes X86 instructions on a native VLIW microprocessor using a software layer, called Code Morphing Software (CMS). Each instruction (molecule) of the VLIW processor can issue two or four RISC-like operations (atoms) to the functional units (FUs). Five FUs are available: two ALUs, a floating point unit, one memory and one branch units. Moreover, the processor has a set of 64 integer and 32 floating point registers.

The CMS is comprised of an interpreter, a dynamic binary translator, an optimizer, and a runtime system. The CMS has several objectives to accomplish so it can execute x86 instructions correctly. Besides implementing the whole instruction set and the architectural registers, it must handle memory mapped I/O and complete exception behavior. Furthermore, the system must be prepared to execute any kind of OS, so it cannot be tailored to a specific one. CMS must be able even to execute BIOS code. Moreover, as the x86 is a general purpose system, the CMS also has to execute with a satisfactory performance diverse programs with different behaviors, and handle things such as self-modifying code.

The role of the translator is to decode x86 instructions, select regions for translation, analyze the code and generate the native VLIW code, optimizing and scheduling it. All these tasks make it the most complex component of the Code Morphing System. The optimizer, besides performing architecture specific optimizations, schedules the VLIW code in a trace with a single-entry point and multiple exit ones (so it is not limited to basic blocks boundaries). The CMS has also a runtime system responsible for handling devices, interrupts, exceptions, power management and the translation cache garbage collector.

Figure 4.6 illustrates the translation process. While interpreting x86 instructions, data on execution frequency, branch directions and memory mapped I/O are gathered. When a certain threshold is achieved, then the translator comes to action, producing native VLIW code for that x86 sequence. The translated sequence is stored in the translation cache, and it will be reused next time the same x86 sequence is found, unless that entry in the translation cache was invalidated for some reason. The exit of the branch at the end of each translated block calls a lookup routine (represented by "no chain" in the figure), which can transfer the control either to another translation block or back to the interpreter, so it can continue translating x86 code. If the control is transferred to another piece of translation, the branch operation is modified so it goes directly there. This process is called chaining. This way, frequently executed regions will be executed entirely on the VLIW processor, avoiding some of the delays caused by branch instructions.

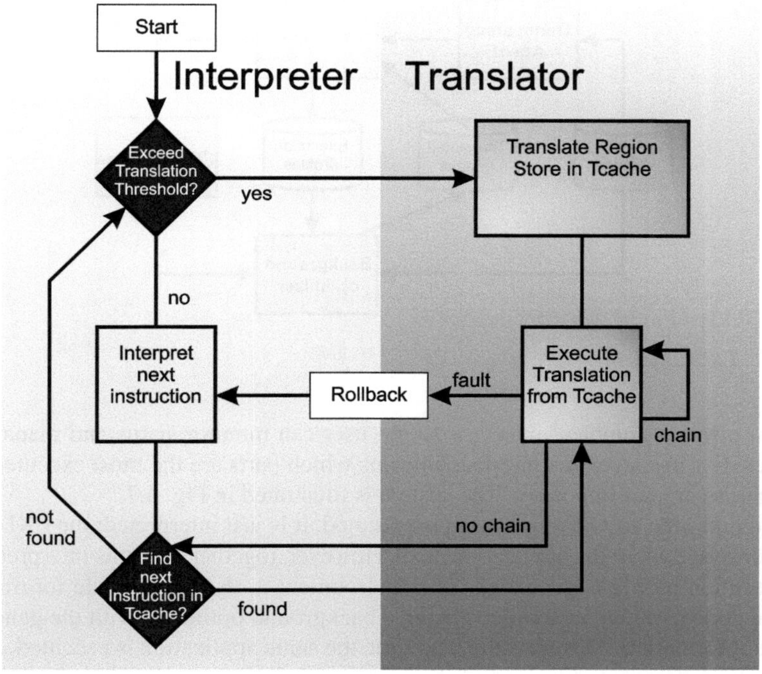

Fig. 4.6 CMS Translation Process

The CMS also supports speculation in the sense that it makes speculative assumptions during the translation, such as the use of memory disambiguation. However, if such assumptions are proven false during execution, they must be handled properly, otherwise incorrect results would be generated. For that, the register set that maps the x86 registers is shadowed, so the VLIW instructions (atoms) work only on the copy of the actual register set. The commit will happen when the execution of that piece of translated code comes to the end. If some exception condition occurs (for instance, a failure in one of the translation assumptions), the runtime system is responsible for rolling back the context to the last commit point (the operations of commit and rollback are also applied to memory operations). Then, starting from that point, the CMS starts interpreting x86 instructions corresponding to that exception, executing the original code in order and treating exceptions when necessary. When exceptions occur repeatedly for a certain chunk of translated code, the CMS generates a more conservative translation trying to diminish the number of exceptions.

4.2.4.5 FX!32

The main goal of FX!32 is to allow transparent execution of native x86 Win32 applications on Alpha Systems [111, 122]. Albeit being transparent to the user, the

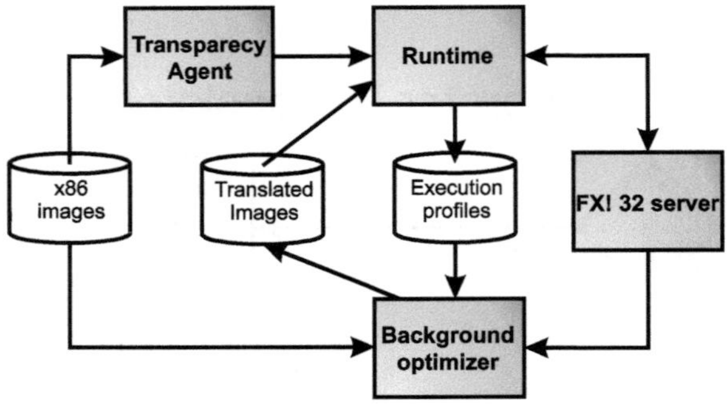

Fig. 4.7 FX!32 System

system offers a graphical interface so the user can monitor status and manage resources. For instance, this interface informs which parts are the most executed, and which ones are not important. The system is illustrated in Fig. 4.7.

The first time an x86 application is executed, it is just interpreted: the FX!32 has no knowledge about the application at all. However, together with this interpretation, an execution profile is generated. The environment is also responsible for translating the code to native instructions using a background optimizer with the generated profile information. This way, the next time the same application is executed, native Alpha code is executed instead of the source x86 for that application. The process of generating native Alpha code is repeated several times until there is stabilization in the profile, so that sufficient performance gains are achieved. According to the authors, this occurs after two or three iterations, indicating that almost all routines were translated. Then, the profile of that image is taken off from the background optimizer list. This way, the first execution of an x86 application will be slow. In the next ones, as the majority of the most used executed code will be already translated to Alpha instructions, performance will improve. It is claimed that, after code translation, gains of up 10 times in performance are achieved when comparing to simple interpretation.

These translated code parts remain in a database provided by the FX!32 system, so they can be accessed next time the x86 application is executed. Translated images are standard DLLs, which are loaded by the Alpha native OS loader (Windows NT). Execution of translated and non-translated code co-exist. While the translated code is executed directly, non-translated is interpreted and profiled (for future translation, if the system decides it is worth to), as already explained. A number of transformations must be done in the code to guarantee correctness of execution between both codes. For instance, the way routines are called is different comparing x86 and Alpha systems (the former use the stack while Alpha uses registers for parameter passing). The translator acts on larger units than basic blocks. According to the authors, the granularity of the translation approximates to the size and structure of a routine.

The server is responsible for coordinating the interface and actions of both interpreter and optimizer. Certain parameters of the server can be configured by the user.

After an x86 program finishes its execution and is unloaded, the server is responsible for merging the new profile information with the old one. Maybe certain parts of code that were not analyzed before require further optimizations.

Launching an X86 application is responsibility of the transparency agent. If there is a call specifying that the process is based on x86 instructions, the transparent agent invokes the FX!32 environment to execute that image. The FX!32 can be executed without any kind of special privileges (although they are necessary when installing the system).

4.3 Reuse

This section discusses dynamic techniques that rely on a very basic principle: parts of the code are repeated during the lifetime of a program execution. Taking advantage of that, such approaches somehow cache the results of previous operations (that can be instructions, basic blocks, or traces) so they can be used again the next time they are found. The next subsections go into details about such techniques, presenting their principles, basic functioning, performance potential, and discussing the viability of actual hardware implementation.

4.3.1 Instruction Reuse

The instruction reuse approach [132–134] relies on the concept of repetition. In [133], the idea of repetition is better clarified. The most general case of repetition is when there are instances of an instruction that appear several times during the program execution with the same input operands and always generating the same result. However, there are cases that the result will be the same even if the input operands are not: a compare instruction can present such behavior. On the other hand, a load instruction can give as result a different value even if its input operands are repeated through time. The Instruction reuse technique only considers the simplest and general case, which is the first case commented above.

Still according to the same article, there are three potential sources of instruction repeatability. The first one is a consequence of the repetition of the input data being processed by a given program. For instance, programs that manipulate text can find repeated sequences of characters (such as spaces or words) during the processing. The same can be considered for programs that work on images. Loops and functions are the second source. For example, instructions that belong to a given loop are constantly repeated even if the processed data is different between multiple interactions: a counter always must be incremented, and a comparison is always performed in order to figure if the loop must continue or not, etc. Finally, there are data structures found in the program that involve repeated accesses to their elements, meaning repeated processing.

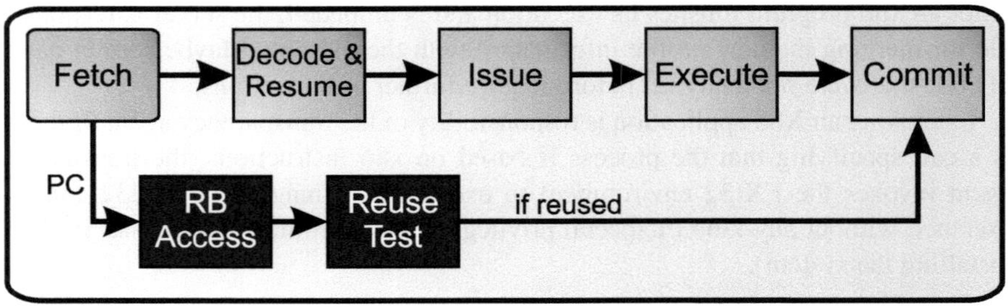

Fig. 4.8 Instruction reuse in a typical processor

The main principle of instruction reuse is that if an instruction with the same operands is repeated a large number of times during the execution of a program, instead of executing it again using a functional unit, the result of this instruction is fetched from a special memory. This result was obtained from previous computations of the very same instruction. In more details, it works as follows: the first time a given instruction is executed, its results are stored in the Reuse Buffer (RB). The entry is indexed by its Program Counter (PC). When the same PC is found again, the entry indexed by that PC is fetched from the RB. This is done while the actual instruction is fetched from memory. The results of the reused instructions will only be written back after passing the validity test done during the decode stage. This test makes sure that the actual instruction would have the same input operands as the one which result was read from the RB, since at the time the RB was accessed, the actual input operands for the current incoming instruction were not ready yet. According to the authors, such approach can be conservative in the sense that if input operands of a given instruction are not ready at the time of reuse verification, then they will not be reused at all. As already commented, instructions that can produce the same result even with different inputs, such as loads, are also not reused. A typical pipeline with instruction reuse hardware is shown in Fig. 4.8.

There are advantages when using this technique: instructions with larger delays (such as multiplications) can be completed faster; data dependent instructions can be reused at the same time; and the path of reused instructions has two pipeline stages less than if they were executed on the processor. Additionally, there are secondary positive effects regarding processor resources, such as freeing functional units, slots in the reservation stations and in the reorder buffer, the reduction of the instruction fetch and data bandwidth (fewer accesses in the register bank and in the memory). These effects potentially increase the possibility of executing additional instructions, if there is still ILP available.

4.3.2 Value Prediction

Value prediction (VP) [117, 118, 125] is very similar to instruction reuse. However, the technique is speculative in the sense that it predicts values for the input operands

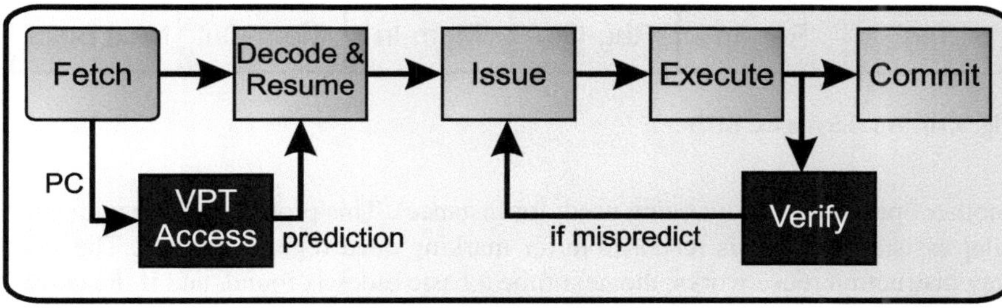

Fig. 4.9 Value prediction

before they are ready. The second difference is that instructions predicted need to be actually executed later so the prediction can be verified (instructions reused are never really executed). Figure 4.9 illustrates VP. The key difference between them is that IR verifies the validity of the results before using them (early validation), while VP uses the results speculatively and verifies them later (late validation). Due to these differences, the two techniques vary in the amount of redundancy they can capture, and the way they interact with other microarchitectural features.

The value prediction technique can be viewed as an extension of the work on load value prediction (LVP) [126], where just that type of instructions was considered.

The predictions are obtained from the Value Prediction Table (VPT), which is implemented in hardware. As the input values are predicted, results can become available before inputs are ready. When the correct values become available later, after actual instruction execution, they are compared to the speculated results. If they are wrong, instructions that used the wrong operands must be re-executed. If the speculation was right, nothing is done and those instructions were executed earlier than they should, bringing performance advantages. The same way it happens with instruction reuse, VP allows data dependent instructions to be completed at the same time, potentially increasing the ILP. Usually IR reduces resource contention (functional units, data cache etc), while on the other hand, VP always increases contention, since predicted instructions always need to be re-executed.

4.3.3 Block Reuse

In [123, 124] another technique is presented, with the purpose of reusing basic blocks (a sequence of instructions with a single entry and single exit points). The authors investigated the input and output values of the basic blocks (BB) and found that they could extend the value prediction and instruction reuse techniques to a coarser granularity, so the performance gains would be larger.

For each basic block, there are the upward-exposed inputs, which are the values that a given basic block will compute, and the live outputs, which are the results that will be actually used in the future (they will not be produced and be written by

Tag	Reg-In	Reg-Out	Mem-In	Mem-Out	Next Block

Fig. 4.10 An entry in the BHB

another operation without being used, for instance). This process is done with compiler assistance: GCC is responsible for marking dead register outputs. The same way instruction reuse works, the next time a basic block is found, and if the current input operands for that basic block are the same as the previous execution, the execution of the whole set of instructions of that basic block is skipped and the live output values are fetched. The input operands are composed of register and memory references. The basic block boundaries are determined at run-time during program execution. An entry point is any instruction after a branch, subroutine call or return. A branch instruction marks an exit point. If an entry point is identified in the middle of a previously found basic block, it is split it two separate basic blocks. The reuse information is stored in the Block History Buffer (BHB). As can be observed in Fig. 4.10, each BHB entry contains 6 fields. The Tag stores the starting address of a basic block. The Reg-In field is divided in several subfields, used to maintain the register references and values of the input context. The Reg-Out has the same purpose, but for the output context. Mem-In and Mem-Out fields are also divided in subentries. Each subentry contains the program counter relative to the instruction that makes the memory access, the memory address, the actual data value and a bit indicating if that entry is being used or not. Finally, the Next Block field is responsible for keeping the address of the basic block that follows the current, if there is also an entry for it in the BHB. The Simplescalar Toolset [110] was employed for this case study. It simulates a MIPS-like processor, using a configuration with four integer ALUs, one integer multiply/divide unit, and the same number of functional units for floating points computation. It is capable of issuing and committing up to four instructions per cycle. The resulting speedup values range from 1.01 to 1.37, with an average of 1.15. The benchmarks were compiled with the GCC O2 level of optimization.

4.3.4 Trace Reuse

The idea of trace reuse [119] extends the previous approach, in the sense that it is applied to a group of instructions (called of trace by the authors), as illustrated in Fig. 4.11. In fact, the authors classify Basic Block Reuse as a particular case of trace-level reuse. Trace is more general, since it can exploit larger sequences of instructions, such as entire subroutines or complex loops.

As the other techniques, trace reuse is based on the input and output contexts. A context is composed of the program counter, registers and memory addresses. Trace reuse works as follows: for a given sequence of instructions, the context of the processor, considering the first instruction of this sequence, is saved. Then, the

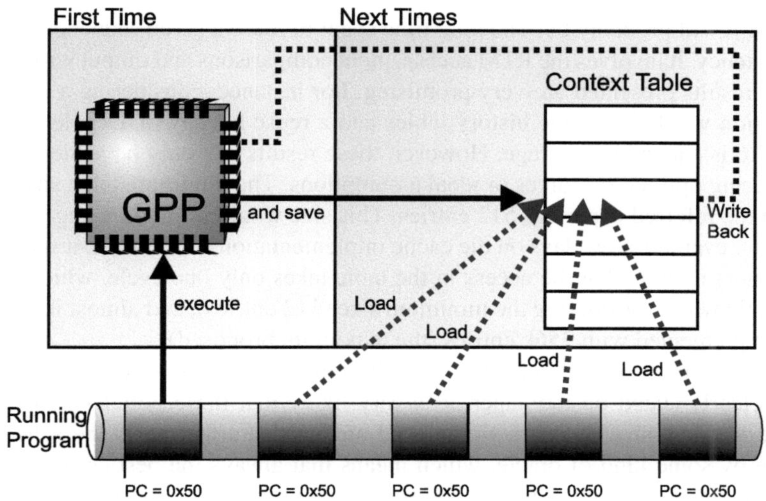

Fig. 4.11 The trace reuse approach

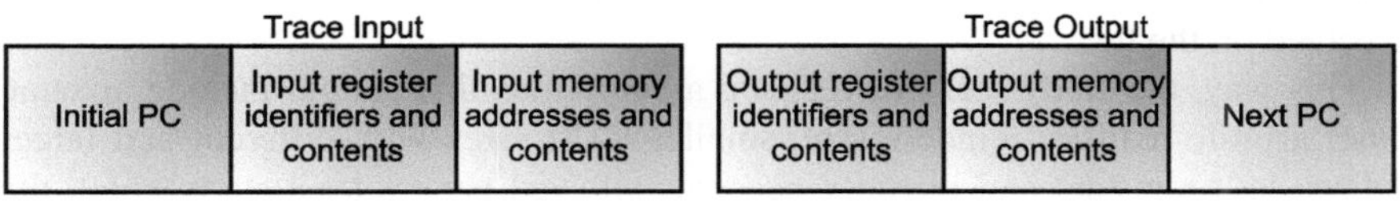

Fig. 4.12 A RTM entry

output context, which is the result of the whole set of instructions that belongs to that sequence, is also saved, after this sequence was normally executed by the processor. After that, each time that the first instruction of this sequence is sent for execution again, the processor state is updated with the output context fetched from a special memory, avoiding the execution of that trace on the processor. This memory is called Reuse Trace Memory (RTM).

Each entry of the RTM is illustrated in Fig. 4.12. These entries are indexed by the PC. Besides the initial PC, each RTM entry is composed of register and memory references and their contents before and after trace execution. Each entry has also the next PC, pointing to the next instruction that should be executed after the trace was reused.

The process of finding which traces should be reused is done dynamically while program is executed. One proposed approach for identifying a trace is to classify instructions as reusable or not. The trace finishes when a non-reusable instruction is found and begins soon after a reusable one is encountered. Another possibility is to use fixed length traces. The size of the trace does not directly influences on the RTM entry, since the instructions themselves are not saved, so the trace size cannot be considered as a limitation. Each trace reuse operation has a latency (since the RTM must be accessed). This way, as more instructions a trace has, the more efficient it will probably be. The total time spent to reuse a given sequence is called reuse latency. It involves the RTM access, input comparisons and output write backs.

The results presented are very promising. For instance, considering a 256-entry instruction window, infinite history tables and a reuse latency of 1 cycle, a speed-up of 3.6 is shown, on average. However, these results can only be achieved when considering optimal resources or ideal assumptions. The minimum table size evaluated in the referred paper has 512 entries. This size would imply in a huge memory footprint, even for nowadays on die cache implementations. Moreover, it seems that the authors assume that the access in the table takes only one cycle, which is very optimistic when considering the minimum size (512 entries), and almost impossible to be implemented with 256k entries (the maximum proposed).

The authors also implemented three different scheduling policies. Although it is not clearly stated on the paper, it is very likely that these policies consider an infinite window size of instructions to be analyzed. Furthermore, the scheduling is done by some kind of oracle, which means that always the best composition of traces is considered to be saved in the special memory. It is important to stand out that defining the best policy for scheduling these instructions can be a very complex job to be done: multiples instructions can compose multiple traces and finding the best combination demands a huge computational effort, which is very hard to be executed on the fly.

This way, the study lacks of realistic assumptions that should include at same time: a finite realistic window size, smaller RTM sizes with different and larger delays, less registers and memory accesses allowed per cycle; a study about the costs of the scheduling algorithm using a finite window; the costs of comparing registers and memory values with the current trace context etc.

4.3.5 Dynamic Trace Memoization and RST

In [112] the authors presented a technique called Dynamic Trace Memoization, which uses memoization tables in order to detect, at real time, traces that can be potentially reused. The technique is very similar to trace reuse, however, it presents a more detailed mechanism on how to perform the job, and an analysis on the hardware costs and different algorithms to perform the task. Traces are built from redundant sequences of instructions. Two tables are used for the memoization mechanism: Memo Table G (Global Memoization Table), responsible for keeping isolated instructions; and Memo Table T (Trace Memoization Table), which holds traces.

The entries in the Memo Table G have the PC address of the instruction, operand values, instruction identification (such as the one informing if it is a branch or jump), among others. Each dynamic instruction can be classified in two ways: redundant or not. Non-supported instructions such as loads and stores are tagged as non-redundant.

For the current incoming instruction, a search is done in the Memo Table G in order to figure if a match occurs. If it is positive, a comparison with the input operators is done. If the input operands are the same, the instruction is tagged as redundant. Otherwise, a new entry for that instruction and input operands is created in the table, and the instruction is classified as non-redundant. The process is repeated until a non-redundant instruction is found. From this information, an entry in the Memo Table T is created. An entry in that table has information such as initial and final PCs, input and output operands etc. A Memo Table T may have multiple instances of the same traces but with a different input contexts.

In [128] this approach was extended in order to support speculative execution. In [127] the technique is combined with value prediction and restricted hardware resources, reducing the number of trace candidates and the size of their contexts, achieving a speedup of 1.21, on average. The basic concept of this approach (RST - Reuse through Speculation on Traces) is to speculate inputs values of a trace when they are not available at the time the trace needs to be reused. This way, RST combines both value prediction and trace reuse approaches. When a trace is speculatively reused, the output values are sent to the commit stage. Therefore, Dispatch, Issue and Execution stages are bypassed for the whole sequence of instructions that compose that trace. Hence, by using RST one can alleviate the pressure on other system resources, such as functional units. As the study extended the work done about DTM, the same Memo Tables G and T were employed. Various experiments were performed: trace reuse with and without speculation, and DTM supporting loads and stores. As in DTM, it is possible to have more than one entry for a given trace. However, it has to be decided which one will be chosen for reuse and speculation. The selection is based on an LRU list.

In this chapter, dynamic techniques used in the GPP market have been presented. In the sequel, how these techniques can be used with the reconfigurable fabric are discussed, so that it can adapt itself to cope with different kernels at runtime, while still sustaining binary compatibility.

References

107. Altman, E.R., Ebcioğlu, K., Gschwind, M., Sathaye, S.: Advances and future challenges in binary—translation and optimization. In: Proc. of the IEEE, pp. 1710–1722 (2001)
108. Altman, E.R., Kaeli, D.R., Sheffer, Y.: Welcome to the opportunities of binary translation. IEEE Comput. **33**(3), 40–45 (2000)
109. Bala, V., Duesterwald, E., Banerjia, S.: Dynamo: a transparent dynamic optimization system. In: PLDI'00: Proceedings of the ACM SIGPLAN 2000 Conference on Programming Language Design and Implementation, pp. 1–12. ACM, New York (2000). doi:10.1145/349299.349303
110. Burger, D., Austin, T.M.: The simplescalar tool set, version 2.0. SIGARCH Comput. Archit. News **25**(3), 13–25 (1997). doi:10.1145/268806.268810
111. Chernoff, A., Herdeg, M., Hookway, R., Reeve, C., Rubin, N., Tye, T., Bharadwaj, S., Yates, J.: Fx!32 a profile-directed binary translator. IEEE Micro **18**(2), 56–64 (1998). doi:10.1109/40.671403
112. Costa, A.T.D., Franca, F.M., Filho, E.M.C.: The dynamic trace memoization reuse technique. In: 9th PACT, 2000, IEEE CS, pp. 92–99 (2000)

113. Dehnert, J.C., Grant, B.K., Banning, J.P., Johnson, R., Kistler, T., Klaiber, A., Mattson, J.: The transmeta code morphing™ software: using speculation, recovery, and adaptive retranslation to address real-life challenges. In: CGO'03: Proceedings of the International Symposium on Code Generation and Optimization, pp. 15–24. IEEE Computer Society, Los Alamitos (2003)
114. Ebcioğlu, K., Altman, E., Gschwind, M., Sathaye, S.: Dynamic binary translation and optimization. IEEE Trans. Comput. **50**(6), 529–548 (2001). doi:10.1109/12.931892
115. Ebcioğlu, K., Fritts, J., Kosonocky, S., Gschwind, M., Altman, E., Kailas, K., Brigh, T.: An eight issue tree-vliw processor for dynamic binary translation. In: ICCD'98: Proceedings of the International Conference on Computer Design, p. 488. IEEE Computer Society, Los Alamitos (1998)
116. Ebcioğlu, K., Altman, E.R.: Daisy: dynamic compilation for 100 architectural compatibility. In: ISCA'97: Proceedings of the 24th Annual International Symposium on Computer Architecture, pp. 26–37. ACM, New York (1997)
117. Gabbay, F., Gabbay, F.: Speculative execution based on value prediction. Tech. Rep., EE Department TR 1080, Technion–Israel Institute of Technology (1996)
118. Gabbay, F., Mendelson, A.: Using value prediction to increase the power of speculative execution hardware. ACM Trans. Comput. Syst. **16**(3), 234–270 (1998). doi:10.1145/290409.290411
119. Gonzalez, A., Tubella, J., Molina, C.: Trace-level reuse. In: ICPP'99: Proceedings of the 1999 International Conference on Parallel Processing, p. 30. IEEE Computer Society, Los Alamitos (1999)
120. Gschwind, M., Ebcioğlu, K., Altman, E., Sathaye, S.: Binary translation and architecture convergence issues for IBM system/390. In: ICS '00: Proceedings of the 14th International Conference on Supercomputing, pp. 336–347. ACM, New York (2000). doi:10.1145/335231.335264
121. Hennessy, J.L., Patterson, D.A.: Computer Architecture: A Quantitative Approach, 4th edn. Morgan Kaufmann, San Mateo (2006)
122. Hookway, R.J., Herdeg, M.A.: Digital fx!32: combining emulation and binary translation. Digit. Tech. J. **9**(1), 3–12 (1997)
123. Huang, J., Lilja, D.: Exploiting basic block value locality with block reuse. In: HPCA'99: Proceedings of the 5th International Symposium on High Performance Computer Architecture, p. 106. IEEE Computer Society, Los Alamitos (1999)
124. Huang, J., Lilja, D.J.: Extending value reuse to basic blocks with compiler support. IEEE Trans. Comput. **49**(4), 331–347 (2000). doi:10.1109/12.844346
125. Lipasti, M.H., Shen, J.P.: Exceeding the dataflow limit via value prediction. In: MICRO 29: Proceedings of the 29th Annual ACM/IEEE International Symposium on Microarchitecture, pp. 226–237. IEEE Computer Society, Los Alamitos (1996)
126. Lipasti, M.H., Wilkerson, C.B., Shen, J.P.: Value locality and load value prediction. In: ASPLOS-VII: Proceedings of the Seventh International Conference on Architectural Support for Programming Languages and Operating Systems, pp. 138–147. ACM, New York (1996). doi:10.1145/237090.237173
127. Pilla, M.L., Childers, B.R., da Costa, A.T., Franca, F.M.G., Navaux, P.O.A.: A speculative trace reuse architecture with reduced hardware requirements. In: SBAC-PAD '06: Proceedings of the 18th International Symposium on Computer Architecture and High Performance Computing, pp. 47–54. IEEE Computer Society, Los Alamitos (2006). doi:10.1109/SBAC-PAD.2006.7
128. Pilla, M.L., da Costa, A.T., França, F.M.G., Childers, B.R., Soffa, M.L.: The limits of speculative trace reuse on deeply pipelined processors. In: SBAC-PAD'03: Proceedings of the 15th Symposium on Computer Architecture and High Performance Computing, p. 36. IEEE Computer Society, Los Alamitos (2003)
129. Sager, D., Group, D.P., Corp, I.: The microarchitecture of the Pentium 4 processor. Intel Technol. J. **1**, 2001 (2001)
130. Shankland, S.: Transmeta shoots for 700 MHz with new chip. In: CNET News (2000)

131. Sites, R.L., Chernoff, A., Kirk, M.B., Marks, M.P., Robinson, S.G.: Binary translation. Commun. ACM **36**(2), 69–81 (1993). doi:10.1145/151220.151227
132. Sodani, A., Sohi, G.S.: Dynamic instruction reuse. SIGARCH Comput. Archit. News **25**(2), 194–205 (1997). doi:10.1145/384286.264200
133. Sodani, A., Sohi, G.S.: An empirical analysis of instruction repetition. SIGOPS Oper. Syst. Rev. **32**(5), 35–45 (1998). doi:10.1145/384265.291016
134. Sodani, A., Sohi, G.S.: Understanding the differences between value prediction and instruction reuse. In: MICRO 31: Proceedings of the 31st Annual ACM/IEEE International Symposium on Microarchitecture, pp. 205–215. IEEE Computer Society, Los Alamitos (1998)
135. Yang, B.S., Moon, S.M., Park, S., Lee, J., Lee, S., Park, J., Chung, Y.C., Kim, S., Ebcioğlu, K., Altman, E.R.: Latte: A java vm just-in-time compiler with fast and efficient register allocation. In: IEEE PACT, pp. 128–138 (1999)

Chapter 5
Dynamic Detection and Reconfiguration

Abstract As very diverse applications have to be executed in the same computational structure, the pressure for dynamic modifications in the reconfigurable logic increases, since fast adaptability is key to sustain fast execution with the lowest possible power dissipation. This proves that the main strategy to bring reconfigurable systems to be used as mainstream computing is to rely on dynamic optimization techniques, such as the ones already presented. Therefore, in this chapter two approaches that use reconfigurable fabric together with a mechanism that somehow reassembles the behavior of the dynamic optimization techniques are discussed, as well as their basic structure, granularity, communication issues, how the binary translation mechanism works and their potential gains in performance and energy.

5.1 Warp Processing

Trying to unify some of the ideas of dynamic optimization with reconfigurable systems, Vahid et al. [143, 144, 146, 150–152] presented the first studies about the benefits and feasibility of dynamic partitioning (which is the process of finding the best parts of software to be executed on hardware) using reconfigurable logic, producing good results for a number of popular embedded system benchmarks. The main purpose of the approach is, just as many others, to use FPGAs to implement the most executed parts of the software, boosting performance. The difference, in this case, is that the speedups are achieved in a totally transparent process, without any user involvement. The system is targeted to speedup specific programs, such as the ones with dataflow behavior and distinct kernels subject of optimization. The software partitioning and decompilation, as well as FPGA synthesis, place and route are done at run time, during the program execution.

This approach, called Warp Processing, is based on a complex SoC. As it can be observed in Fig. 5.1, the system is composed of a microprocessor to execute the application software, another microprocessor where a simplified CAD algorithm runs, local memory and a dedicated FPGA array. As it happens in conventional BT systems described in the previous chapter, the first time an application is executed

A.C. Schneider Beck Fl., L. Carro, *Dynamic Reconfigurable Architectures and Transparent Optimization Techniques*,
DOI 10.1007/978-90-481-3913-2_5, © Springer Science+Business Media B.V. 2010

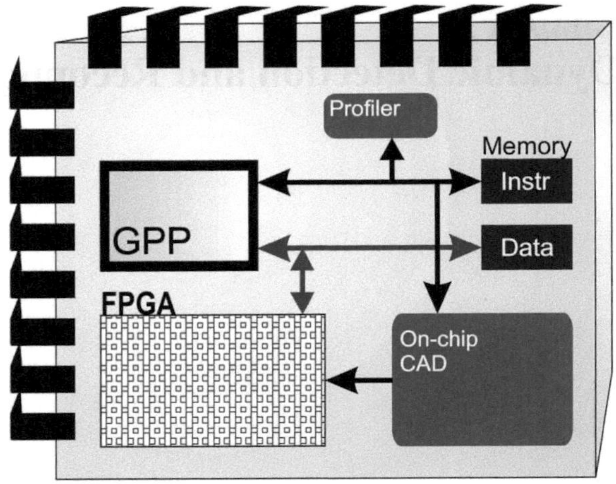

Fig. 5.1 The Warp processor system

on the GPP, a profiler monitors its behavior, so it can find the critical kernels. Then, the on-chip CAD running in the SOC starts working to implement these software parts to be executed as reconfigurable instructions on the FPGA. The profiler is non-intrusive, in the sense that it monitors instructions using the instruction memory bus, being located outside the GPP.

The authors claim that Warp Processing aims to optimize embedded systems that repeatedly execute the same standalone application for extended periods. The importance of backwards compatibility is also highlighted. Moreover, Warp Processing could also be used for applications that execute several times on the system at different periods, as long as there is available a mechanism to save the configurations, so that they could be reused in the future.

5.1.1 The Reconfigurable Array

When comparing to a regular off-the-shelf FPGA, the one implemented for the Warp System is simplified regarding its structure, so technology mapping is simpler, allowing higher clock frequencies. This FPGA, called W-FPGA, is composed of registers and CLBs with two 3-input 2-outputs LUTs each (Fig. 5.2). Moreover, the W-FPGA has a 32-bit MAC (multiplier-accumulator), a DADG (Data Address Generator) to handle memory accesses, and uses a special routing scheme [144].

The registers found in the W-FPGA can be used as input to the MAC or be directly connected to the reconfigurable fabric. In addition, the outputs from the configurable logic are connected to those registers, so that they can store intermediate values during a given computation. The registers are also responsible for interfacing to the rest of the system. The W-FPGA is organized this way because the main purpose of Warp Processing is to optimize loops and DSP like applications. For instance, the computation of loop bounds or sequential memory addresses become

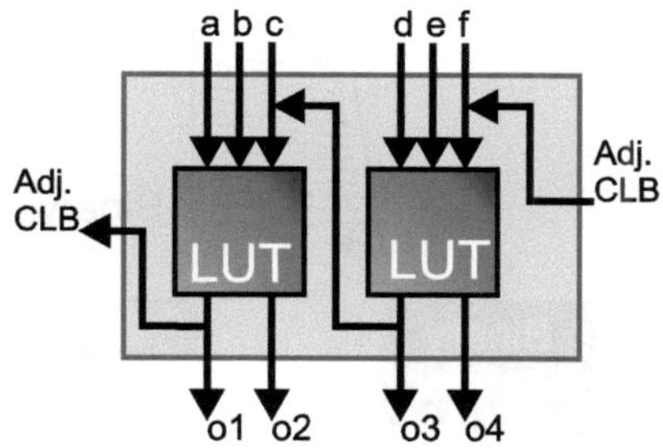

Fig. 5.2 Simplified Warp FPGA structure: CLB with two LUTs

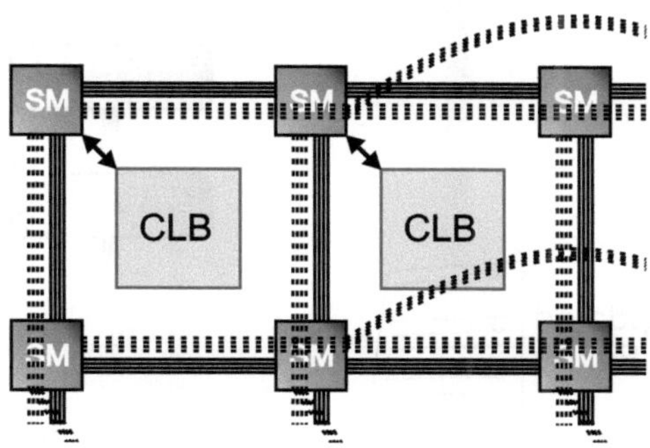

Fig. 5.3 Simplified Warp FPGA structure: routing

faster when using the DADG. MAC, in turn, is a very common operation found in DSP applications. This way, implementing it directly in dedicated hardware instead of using the configurable logic brings huge performance gains.

As can be observed in Fig. 5.3, the CLBs are surrounded by switch matrices, and each CLB is connected to one of them. The switch matrix can route data to the four adjacent switches, or to one switch two rows apart vertically, or to another switch also two columns apart horizontally. Albeit this is a restriction in the routing mechanism, it facilitates the job for the CAD algorithm, making the FPGA cheaper to be implemented. CLBs are also capable of supporting carry chains internally (connecting their LUTs), or externally, being directly connected to adjacent CLBs within the same row. Hence, components such as adders and comparators that need carry logic can be easily implemented regarding the routing mechanism.

5.1.2 How Translation Works

The Translation system is illustrated in Fig. 5.4, and the following steps for its functioning are necessary:

1. Initially, the software binary is loaded into the instruction memory;
2. The microprocessor executes the instructions from this software binary;
3. Profiler monitors the instructions and detects critical regions in binary.

Then, the on-chip CAD:

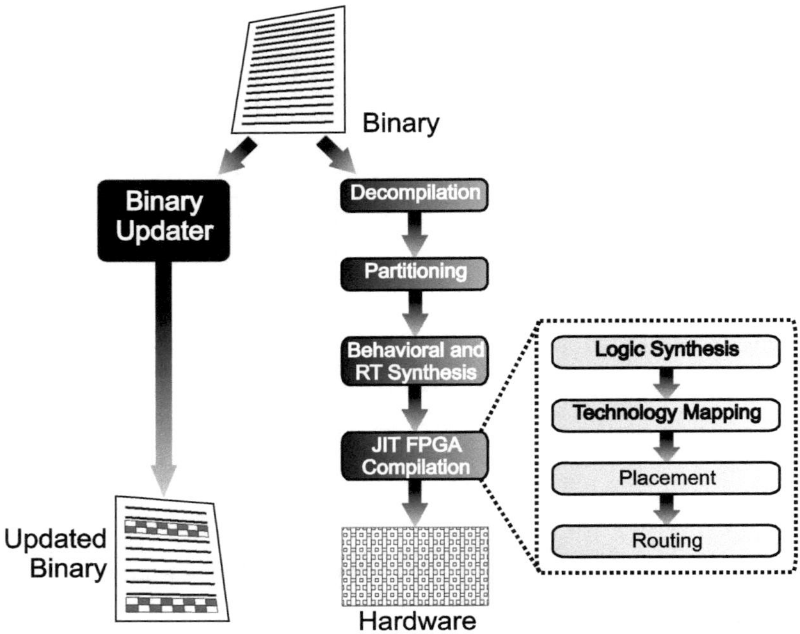

Fig. 5.4 Steps performed by the CAD software

1. reads in critical regions;
2. decompiles a given critical region into a control data flow graph (CDFG);
3. synthesizes the decompiled CDFG to a custom (parallel) circuit;
4. maps this circuit onto FPGA;
5. replaces instructions in the original binary to use the FPGA hardware.

As already explained, the profiler is responsible for finding the hot spots of the binary code. The entry points for parts of code that can be optimized are backward branches. Each time the profiler finds such instructions, it uses a specific cache of branch frequencies. This way, it is possible to identify hot spots that should be optimized. Once a hot spot is found, the on-chip CAD starts working.

The first task of the on-chip CAD is decompilation. It converts the binary code to an intermediate language, which has its own instruction set. Then, both control and dataflow graphs are built. With these graphs, the decompiler performs some compiler optimizations. Two new techniques were proposed: loop rerolling, used to detect unrolled loops and transform them back to their original format, since the gains of warp processing are highly based on loop optimizations; and operator strength promotion, which finds transformed operations, such as multiplications transformed to a sequence of shifts and adds, also translating them back to their original format. This way, it is possible to use the specific circuits available in FPGA to perform the function (in this case, the hardware dedicated multiplier).

After that, it is time to partition the code. Profiler information is used in order to figure which kernels are suitable for hardware implementation. Once the critical regions are identified, the RT synthesis converts the graphs of those regions to a hardware circuit description, which will be later converted to a netlist format, specifying that circuit using Boolean expressions for each output. Then, the Just In Time FPGA compiler starts mapping that netlist to FPGA.

The mapping process involves several phases. The first task of the JIT FPGA compiler is logic synthesis, in order to optimize the circuit. This is done by using an algorithm to minimize an acyclic graph based on the Boolean logic network. In this network, each node corresponds to simple 2-input gates such as ADD, OR etc. With the optimized circuit, technology mapping starts working. It uses a graph clustering algorithm, to combine the nodes of the graph to create 3-input/2-output LUT nodes. Then they can be mapped directly to the W-FPGA. The LUTs are packed into CLBs trying to minimize communication costs. Finally, the CLBs are put onto the configurable logic and their inputs and outputs are routed. According to the authors, routing is the most compute- and memory-intensive CAD task.

The last step is to update the software binary so it can use the just generated hardware parts the next time it is executed. The original instructions are replaced for others to handle the communication and control of the hardware. During hardware execution, the processor is shut down. It is turned on again when a completion signal arrives from the FPGA. If the system figures that the new hardware would result in a slowdown, the binary updater does not change the original program for that part.

5.1.3 Evaluation

In [143], the CAD algorithm was executed on an ARM7, including separate cache and memory (instruction and data). On average, it takes 1.2 seconds to execute on an ARM7 running at 40 MHz. The authors claim that it would be possible to eliminate the extra processor and execute the CAD module together with the regular applications on the same processor, though. The warp processor was compared with several HW/SW partitioning approaches, with performance and energy data. Benchmarks from different sets were considered (such as NetBench [147] and Mediabench [142]).

In [152] it is claimed that the FPGA fabric supports nearly 50,000 equivalent gates. In 0.18-μm technology, the W-FPGA would take approximately the same area as an ARM9 processor with 32 Kb, or the same as a standalone 64-Kbyte of cache memory.

In [145], results show the benefits of warp processing for soft-core processors. The technique was implemented in a MicroBlaze-based FPGA. Several embedded systems applications from the Powerstone and EEMBC benchmark suites were analyzed. The experimental setup considers a MicroBlaze processor implemented using the Spartan3 FPGA. The MicroBlaze processor core has a maximum clock frequency of 85 MHz. However, the remaining FPGA circuits can operate at up to 250 MHz. The processor was configured to include a barrel shifter and multiplier, as the applications considered required both operations. In the same article, they present the performance speedup and energy reduction of the MicroBlaze-based warp processor compared with a standalone MicroBlaze processor. The software application execution was simulated on the MicroBlaze using the Xilinx Microprocessor Debug Engine, where instruction traces for each application were obtained. This trace was used to simulate the behavior of the on-chip profiler to determine the single most critical region within each application.

The system was also compared with readily available hard-core processors. In overall, the MicroBlaze extended with Warp Processing had better performance than the ARM7, ARM9, and ARM10 processors, and requires less energy than the ARM10 and ARM11 processors. The ARM11 processor executing at 550 MHz is on average 260% faster than the MicroBlaze warp processor, but requires 80% more energy. Furthermore, compared with the ARM10 executing at 325 MHz, the MicroBlaze warp processor is on average 30% faster while requiring 26% less energy. Therefore, while the MicroBlaze warp processor is neither the fastest nor the lowest energy alternative, it is comparable and competitive with existing hard-core processors, while having all the flexibility advantages associated with soft-core processors.

5.2 Configurable Compute Array

In [136–140] the Configurable Compute Array (CCA) was proposed. CCA is a coarse grain reconfigurable array tightly coupled to an ARM processor.

5.2.1 The Reconfigurable Array

The proposed CCA is implemented as a matrix of heterogeneous functional units (FUs). There are two types of FUs in this design, referred to as type A and B, for simplicity. Type A FUs perform 32-bit addition/subtraction as well as logical operations. Type B FUs perform only the logical operations, which include and/or/xor/not, sign extension, bit extraction, and moves. To ease the mapping of subgraphs onto the CCA, each row is composed of either type A FUs or type B FUs.

The matrix can be characterized by the depth, width, and operation capabilities. Depth is the maximum length dependence chain that a CCA will support. This corresponds to the potential vertical compression of a dataflow subgraph. Width is the number of FUs that can work in parallel. This represents the maximum instruction-level parallelism (ILP) available to a subgraph execution. Figure 5.5 shows the block diagram of a CCA with depth 7. In this figure, type A functional units (FU) are represented with white squares and type B units with gray squares. The CCA has 4 inputs and 2 outputs. Any of 4 inputs can drive the FUs in the first level. The first output delivers the result from the bottom FU in the CCA, and the second output is optionally driven from an intermediate result from one of the other FUs.

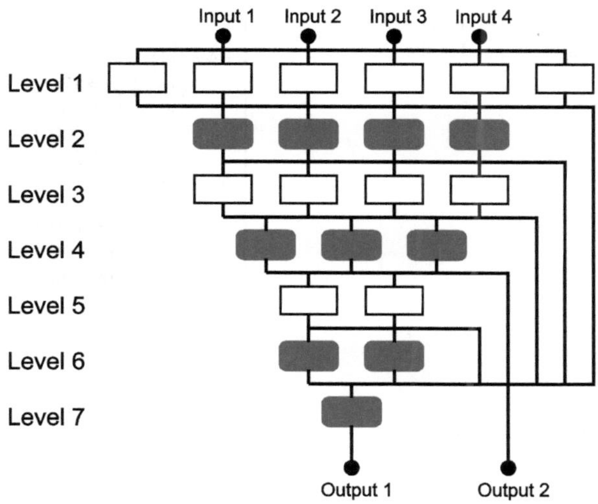

Fig. 5.5 Example of a CCA with 4 inputs and 2 outputs, and depth of 7

5.2.2 Instruction Translator

Feeding the CCA involves two steps: the discovery of which subgraphs are suitable for running on the CCA, and their replacement by microops in the instruction stream. Two alternative approaches are presented: static and dynamic. Static discovery finds subgraphs for the CCA at compile time. Those are marked in the machine code by using two additional instructions, so that a replacement mechanism can insert the appropriate CCA microops dynamically. Using these instructions to mark patterns allows for binary forward compatibility, meaning that as long as future generations of CCAs support at least the same functionality of the one it was compiled for, the subgraphs marked in the binary will still be useful. However, as the code is changed, the backward compatibility is lost anyway.

Dynamic discovery, in turn, assumes the use of a trace cache to perform subgraph discovery on the retiring instruction stream. Its main advantage is that the use of the CCA is completely transparent to the ISA. Theoretically, the static discovery technique can be much more complex than the dynamic version, since it is performed offline; thus, it does a better job on finding subgraphs. Figures 5.6 and 5.7 demonstrate how a sequence of instructions is mapped into a typical CCA configuration. The CFG representing a part of the code (Fig. 5.7) is shown in Fig. 5.6. The bold circles represent the instructions that are in the critical path. These instructions will be mapped to the CCA. Finally, Table 5.1 shows the delays of the functional units that will be used for this sequence. This measurement proves that it is possible to perform more than one single computation within a single clock cycle without affecting the critical path. The shifters do not present any delay because they use compile time constants.

The instruction grouping discovery technique proposed to be used together with the CCU is highly based on the rePlay Framework [148]. The process works as follows: initially, the application is profiled to identify frequently executed kernels,

Fig. 5.6 Part of code that will be mapped to the CCA

```
m ^= p[0];
r ^= p[1];
r ^= ( ( ( s[ (m>>24L) ]+
          s[ 0x0100 + (( m>>16L) & 0xff ]) ^
          s[ 0x0200 + (( m>>  8L) & 0xff ]) +
          s[ 0x0300 + (( m ) & 0xff ]) & 0xffffffff;
m ^= p[2];
m ^= ( ( ( s[ (r >>24L) ]+
          s[ 0x0100 + (( r >>16L) & 0xff ]) ^
          s[ 0x0200 + (( r >>  8L) & 0xff ]) +
          s[ 0x0300 + (( r ) & 0xff ]) & 0xffffffff;
```

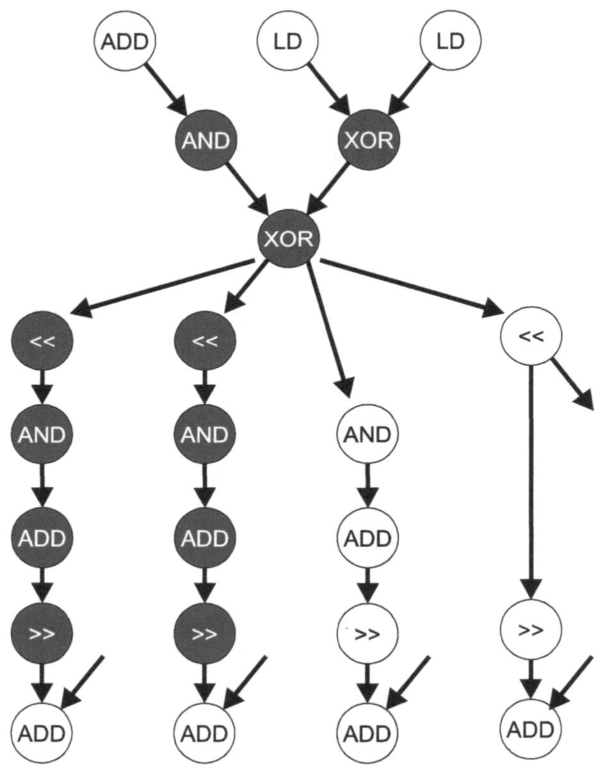

Fig. 5.7 After mapping

Table 5.1 Delays

Opcode	Adders	Cycles
ADD	1.0	0.3
AND	0.12	0.06
Shift	0.01	~ 0
XOR	0.16	0.09

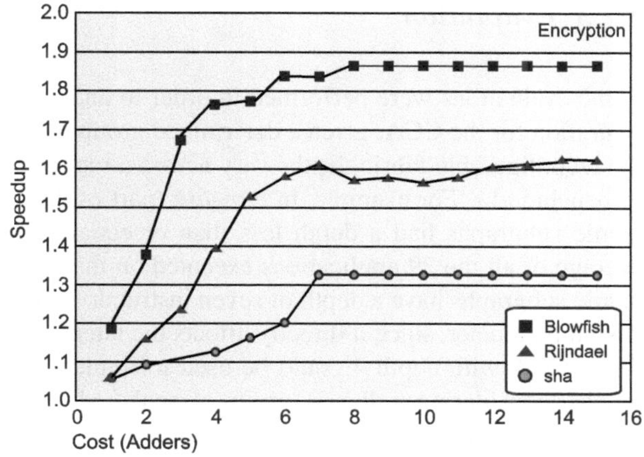

Fig. 5.8 Speed-up versus Area overhead, represented by the cost of adders

called frames. The most frequently executed ones are then analyzed and subgraphs that can be beneficially executed on the CCA are selected. Then, the compiler generates machine code for the application, explicitly identifying the optimized subgraphs to facilitate simple dynamic replacement during execution. Frames have the same purpose of superblocks [141] or use the same principle of trace cache [149]; they have one single entry point and one single exit point, encapsulating one single flow of control in an atomic fashion: if one instruction within a given frame is executed, the rest of the instructions is also executed. A frame is composed of instructions based on speculative branch results. If one transformed branch (assertion) is miss predicted inside the frame, the whole frame execution is discarded.

The subgraphs considered were limited to have at most four inputs and two outputs. Furthermore, memory, branch, and complex arithmetic operations were excluded from the subgraphs. Previous work [153] has shown that allowing more than four input or two output operands would result in very modest performance gains when memory operations are not allowed in subgraphs. In Fig. 5.8 one can observe the potential of implementing a CCA together with the microprocessor, demonstrating the speedup versus a relative area cost of each CCA for three different applications. As it can be seen, with a small cost in terms of hardware, good performance improvements can be achieved.

It is important to point out that operations involving more expensive multiplier/divider circuits are not allowed in subgraphs, because of latency issues. Additionally, memory operations are also disallowed. Load operations have non-uniform latencies, due to cache effects, so supporting them would entail incorporating stall circuitry into the CCA. Although shifts did constitute a significant portion of the operation mix, barrel shifters were too large and incurred too much delay for a viable CCA implementation considering project restrictions.

5.2.3 Evaluation

Some evaluations were performed in order to analyze what would be the best configuration for the CCA, given a determined group of benchmarks. It was shown that the reconfigurable fabric depths vary across a representative subset of three groups of benchmarks. For example, in *blowfish* (part of the MIBench set), 81.42% of dynamic subgraphs had a depth less than or equal to four instructions. Taking the average of all the 29 applications executed on the system, about 99.47% of the dynamic subgraphs have a depth of seven instructions or less. Graph depth is a critical design parameter, since it directly affects the latency of the CCA. It was discovered that a CCA with depth 4 could be used to implement more than 82% of the subgraphs, considering a diverse group of analyzed applications. Going below depth of 4 seriously affects the coverage of subgraphs that can be executed on the CCA. Therefore, only CCAs with depths between 4 and 7 were considered in the study.

The search for the ideal width was also performed. Using the same set of applications, it was figured that 4.2% of dynamic subgraphs had width of 6 or less in row 1, with only 0.25% of them having width 7 of more. In the following rows of the matrix, the widths decrease. For instance, the average width in row 2 is 4 or 5. This data suggests that a CCA should be triangularly shaped to maximize the number of subgraphs supported without wasting area resources.

5.3 Drawbacks

There are some drawbacks when using the aforementioned techniques. Concerning the Warp Processing, the first one is that it uses a complete SOC, with different hardware communicating with each other, which could increase the design cycle time and make it harder to test. Moreover, even if the CAD system used is simplified, it remains complex: it does decompilation, CFG analysis, place and route etc, requiring significant resources: up to 8 MB of memory are necessary for its execution, still big and power hungry for nowadays on-die memories. Another deficiency is related to the FPGA: besides the long latency and area overhead, it is also power inefficient, due to the excessive switches and the considerable amount of static power dissipated. Moreover, because of the memory footprint required for storing configurations, this technique is just limited to critical parts of the software, working at its best just in very particular programs, such as the filter based ones, being very restrictive even when considering embedded systems. Floating point arithmetic, dynamic memory allocation, recursion and the user of pointers other than array accesses are not allowed.

The CCA does not support memory operations, shifts and multiplications—or any operation that involves a different delay when comparing to the functional units employed, limiting its field of application. Therefore, the context has a limited number of inputs and outputs. Moreover, it uses very complicated graph analysis and changes the binary in the static discovery. In the same way, the dynamic approach also makes use of a complex graph analysis, based on RePlay [148], which leads to a huge memory overhead. Because of that, just high-level simulations using the Simplescalar Toolset are reported. No measurements are given in terms of area overhead, power consumption and timing and there are no details about how a CGF is transformed to an array's configuration at run time. The overheads considering the array, and the detection and reconfiguration delays are not discussed at all.

Despite all the aforementioned drawbacks, both works discussed previously are very important, because they show the potential of executing parts of the software in reconfigurable logic and its feasibility.

References

136. Clark, N., Blome, J., Chu, M., Mahlke, S., Biles, S., Flautner, K.: An architecture framework for transparent instruction set customization in embedded processors. In: ISCA '05: Proceedings of the 32nd Annual International Symposium on Computer Architecture, pp. 272–283. IEEE Computer Society, Los Alamitos (2005). doi:10.1109/ISCA.2005.9
137. Clark, N., Kudlur, M., Park, H., Mahlke, S., Flautner, K.: Application-specific processing on a general-purpose core via transparent instruction set customization. In: MICRO 37: Proceedings of the 37th Annual IEEE/ACM International Symposium on Microarchitecture, pp. 30–40. IEEE Computer Society, Los Alamitos (2004). doi:10.1109/MICRO.2004.5
138. Clark, N., Tang, W., Mahlke, S.: Automatically generating custom instruction set extensions. In: Workshop on Application-Specific Processors (WASP), pp. 94–101 (2002)
139. Clark, N., Zhong, H., Mahlke, S.: Processor acceleration through automated instruction set customization. In: MICRO 36: Proceedings of the 36th Annual IEEE/ACM International Symposium on Microarchitecture, p. 129. IEEE Computer Society, Los Alamitos (2003)
140. Clark, N.T., Zhong, H.: Automated custom instruction generation for domain-specific processor acceleration. IEEE Trans. Comput. **54**(10), 1258–1270 (2005). doi:10.1109/TC.2005.156. Member-Mahlke, Scott A.
141. Hwu, W.M.W., Mahlke, S.A., Chen, W.Y., Chang, P.P., Warter, N.J., Bringmann, R.A., Quellette, R.G., Hank, R.E., Kiyohara, T., Haab, G.E., Holm, J.G., Lavery, D.M.: The superblock: an effective technique for vliw and superscalar compilation. In: Instruction-level Parallel Processors, pp. 234–253 (1995)
142. Lee, C., Potkonjak, M., Mangione-smith, W.H.: Mediabench: A tool for evaluating and synthesizing multimedia and communications systems. In: International Symposium on Microarchitecture, pp. 330–335 (1997)
143. Lysecky, R., Stitt, G., Vahid, F.: Warp processors. ACM Trans. Des. Autom. Electron. Syst. **11**(3), 659–681 (2006). doi:10.1145/1142980.1142986
144. Lysecky, R., Vahid, F.: A configurable logic architecture for dynamic hardware/software partitioning. In: DATE '04: Proceedings of the Conference on Design, Automation and Test in Europe, p. 10480. IEEE Computer Society, Los Alamitos (2004)
145. Lysecky, R., Vahid, F.: A study of the speedups and competitiveness of fpga soft processor cores using dynamic hardware/software partitioning. In: DATE '05: Proceedings of the Conference on Design, Automation and Test in Europe, pp. 18–23. IEEE Computer Society, Los Alamitos (2005). doi:10.1109/DATE.2005.38
146. Lysecky, R., Vahid, F.: Design and implementation of a MicroBlaze-based warp processor. ACM Trans. Embed. Comput. Syst. **8**(3), 1–22 (2009). doi:10.1145/1509288.1509294
147. Memik, G., Mangione-Smith, W.H., Hu, W.: NetBench: a benchmarking suite for network processors. In: ICCAD '01: Proceedings of the 2001 IEEE/ACM International Conference on Computer-aided Design, pp. 39–42. IEEE Press, New York (2001)

148. Patel, S.J., Lumetta, S.S.: Replay: A hardware framework for dynamic optimization. IEEE Trans. Comput. **50**(6), 590–608 (2001). doi:10.1109/12.931895
149. Rotenberg, E., Bennett, S., Smith, J.E.: Trace cache: a low latency approach to high bandwidth instruction fetching. In: MICRO 29: Proceedings of the 29th Annual ACM/IEEE International Symposium on Microarchitecture, pp. 24–35. IEEE Computer Society, Los Alamitos (1996)
150. Stitt, G., Lysecky, R., Vahid, F.: Dynamic hardware/software partitioning: a first approach. In: DAC '03: Proceedings of the 40th Annual Design Automation Conference, pp. 250–255. ACM, New York (2003). doi:10.1145/775832.775896
151. Stitt, G., Vahid, F., McGregor, G., Einloth, B.: Hardware/software partitioning of software binaries: a case study of h.264 decode. In: CODES+ISSS '05: Proceedings of the 3rd IEEE/ACM/IFIP International Conference on Hardware/Software Codesign and System Synthesis, pp. 285–290. ACM, New York (2005). doi:10.1145/1084834.1084905
152. Vahid, F., Stitt, G., Lysecky, R.: Warp processing: Dynamic translation of binaries to fpga circuits. Computer **41**(7), 40–46 (2008). doi:10.1109/MC.2008.240
153. Yu, P., Mitra, T.: Characterizing embedded applications for instruction-set extensible processors. In: DAC '04: Proceedings of the 41st Annual Design Automation Conference, pp. 723–728. ACM, New York (2004). doi:10.1145/996566.996764

Chapter 6
The DIM Reconfigurable System

Abstract As it was shown throughout this book, in order to use reconfigurable computing across several application domains, one must ensure, at the same time, dynamic configuration (to adapt to different program characteristics), software compatibility (since one cannot discard the issue of software legacy, always a pressing issue), and energy efficiency (due to the limits of integration or to cover mobile markets). In this chapter, a reconfigurable machine that covers all these aspects is discussed.

6.1 Introduction

While the number of embedded systems is growing, a new trend can be observed: the presence of multi-functional devices, which perform a wide range of different applications with diverse behaviors, e.g. present day portable phones or PDAs. Therefore, simple processors are not enough to handle the computational requirements of these new systems anymore, thus forcing designers to create novel solutions to increase their performance, while maintaining power dissipation as low as possible. Furthermore, new marketing strategies have been focusing on increasing the device functionalities during the product life cycle to reach a wider market.

As an example, Iphone uses a system on a chip (SoC) to execute its many available functions. A general-purpose ARM processor is responsible for the management of several other processing elements like 3D-graphics coprocessor, audio and video. Moreover, common desktop processors techniques running at almost 700 MHz are found in this platform: SIMD execution, ARM Jazelle and DSP extensions. Basically, each technology that was incorporated in this architecture attacks a specific application domain. For instance, SIMD execution aims to provide high performance on integer multimedia applications, while the Jazelle works as a Java acceleration engine. This scenario illustrates a current embedded design strategy, which is based on several different circuits, each one built to perform a specific function for a defined application domain.

A.C. Schneider Beck Fl., L. Carro, *Dynamic Reconfigurable Architectures and Transparent Optimization Techniques*,
DOI 10.1007/978-90-481-3913-2_6,

In the general-purpose computation market, the application heterogeneity is an order of magnitude higher and there is always a need to reach better performance speedups. As discussed in the begging of this book, superscalar processors are reaching their limits when it comes to accelerate single thread software. Thus, the extra area available has been employed to include extra chips in the die. However, there are programs that are hard to parallelize. Furthermore, some of them need recompilation to benefit from a multithread hardware, meaning that all the issues concerning code recompilation discussed in the Binary Translation section emerge again. In the same way, applications with a high rate of communication and small pieces of data that must be transferred may be not worth to be distributed. Consequently, considering a wide range of applications that are executed on the computer, there is always the need of speeding up a single thread. In some fields with massively distributed physical characteristics (time forecast, atomic simulations, earthquake predictions, etc.), the need for high performance computing is even larger.

Nevertheless, there is also the necessity for power savings. While in mobile embedded systems the main need is to sustain the device working as long as possible without recharging the battery, in the general and high performance computation the worries are focused on overheating issues.

As already discussed and somehow demonstrated, reconfigurable systems are potential candidates to be an architectural alternative to handle these issues. However, they have two main drawbacks. The first one is that they are designed to handle very data intensive or streaming workloads. This means that the main design strategy is to consider the target applications as having very distinct kernels for optimization. By speeding up small parts of the software, huge gains would be achieved. In contrast, as commented before, the number of applications running on a system (embedded or desktop) is growing. The second problem is that the process of mapping pieces of code to reconfigurable logic usually involves some kind of transformation, manual or using special languages or tool chains. These transformations modify somehow the source or the binary code, precluding the wide spread usage of reconfigurable systems. As already shown, sustaining binary compatibility, allowing legacy code reuse and traditional programming paradigms are key factors to reduce the design cycle and maintain backward compatibility.

While the approaches presented in the previous chapter handle some of the aforementioned issues, they cannot attack properly both at the same time. This way, the approach here discussed has several objectives: accelerate any kind of system, no matter its behavior; be totally transparent, meaning that no modifications in the binary code is needed at all, and consequently guaranteeing backward and forward compatibility; and to be implementable in hardware considering nowadays technologies. In the following sub-sections the structure of the system and the case studies are demonstrated. In addition, a quantitative analysis is performed, so that the potential gains of using such system are shown. A detailed analysis in power and energy consumption, area overhead and performance speedups are presented.

6.1.1 General System Overview

The DIM reconfigurable system [158, 159, 161–163] is a set of two different mechanisms: the first one, which is the reconfigurable array, and the second, a binary translation algorithm implemented in hardware. The special BT hardware is called Dynamic Instruction Merging (DIM). The basic system functionality is illustrated in Fig. 6.1. DIM is designed to detect and transform instruction groups for reconfigurable hardware execution. This is done concurrently while the main processor fetches other instructions. When a sequence of instructions is found, following given policies that will be explained later, a binary translation is applied to it. Thereafter, this configuration is saved in a special cache, and indexed by the program counter (PC).

The next time the saved sequence is found, the dependence analysis is no longer necessary: the processor loads the previously stored configuration from the special cache, the operands from the register bank, and activate the reconfigurable hardware as functional unit. Then, the array executes that configuration in hardware (including write back of the results), instead of normal processor instructions. Finally, the PC is updated, in order to continue with the execution of the normal (not translated) block of instructions. This way, repetitive dependence analysis for the same sequence of instructions is avoided. Depending on the size of the special cache used to store the configurations, the optimization can be extended to the entire application, not being limited to very few hot spots. Moreover, both the DIM engine and the reconfigurable

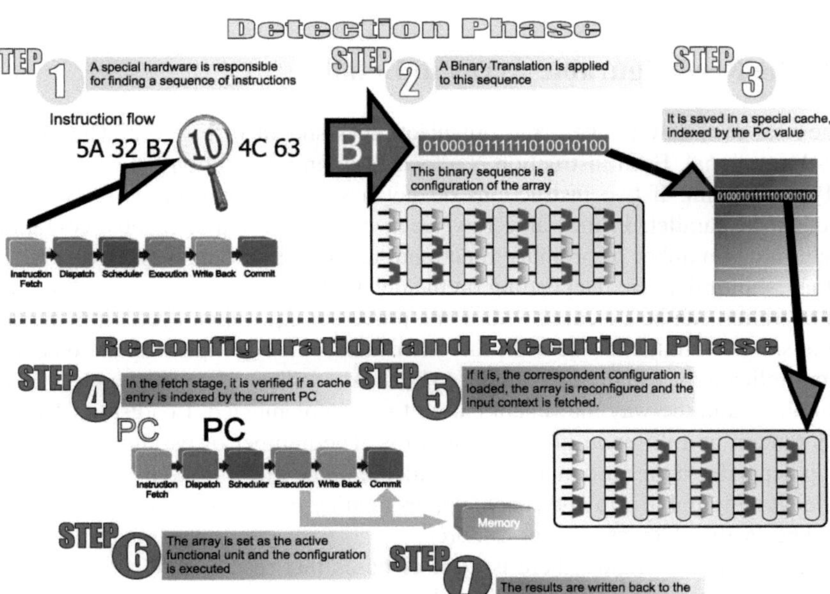

Fig. 6.1 How the DIM system works

array are designed to work in parallel to the processor and do not introduce any delay overhead or penalties in the critical path of the pipeline structure.

The reconfigurable array is tightly coupled to the processor, working as another ordinary functional unit in the pipeline. It is composed of ordinary arithmetic functional units, as ALUs and multipliers, to perform the computation. A set of multiplexers is responsible for the routing. The use of a coarse grain array makes the job of the DIM algorithm easier: since it has a small context size and less complexity in its structure, it becomes more suitable for this kind of dynamic technique.

By using binary translation to avoid source code recompilation or the utilization of extra tools, the optimization process is totally transparent to the programmer. Consequently, such approach does not require extra designer effort and causes no disruption to the standard tool flow used during the software development. Furthermore, depending on the size of the special cache used to store the configurations, the optimization can be extended to the entire application, not being limited to very few hot spots.

Comparing to the techniques cited in the previous chapter, because of the use of a coarse grain unit, it can be implemented in any technology, not being limited to FPGAs only. Adding to the fact that the array is not limited to the complexity of fine-grain configurations, the binary code detection and translation algorithm is very simple and supports any kind of integer instructions, including memory accesses. It can be implemented using trivial hardware resources, in contrast to the complex on-chip CAD software or graph analyzers employed by the approaches mentioned in Chap. 5.

6.2 The Reconfigurable Array in Details

A general overview of the array organization is shown in Fig. 6.2. The array is two-dimensional. Each instruction is allocated in an intersection between one row and one column. If two instructions do not have data dependences, they can be executed in parallel, in the same row. Each column is homogeneous, containing a determined number of ordinary functional units of a particular type, e.g. ALUs, shifters, multipliers etc. Depending on the delay of each functional unit, more than one operation can be executed within one processor equivalent cycle. It is the case of the simple arithmetic ones. On the other hand, more complex operations, such as multiplications, usually take longer to be finished. The delay is dependent on the technology and the way the functional unit was implemented. Load/store (LD/ST) units remain in a different group of the array. The number of parallel units in this group depends on the amount of ports available in the memory. The current version of the reconfigurable array does not support floating point operations, although this could be easily added as another resource.

For the input operands, there is a set of buses that receive the values from the registers. These buses will be connected to each functional unit, and a multiplexer is responsible for choosing the correct value. As can be observed in more details in Fig. 6.3, for each FU, there are two multiplexers that will make the selection

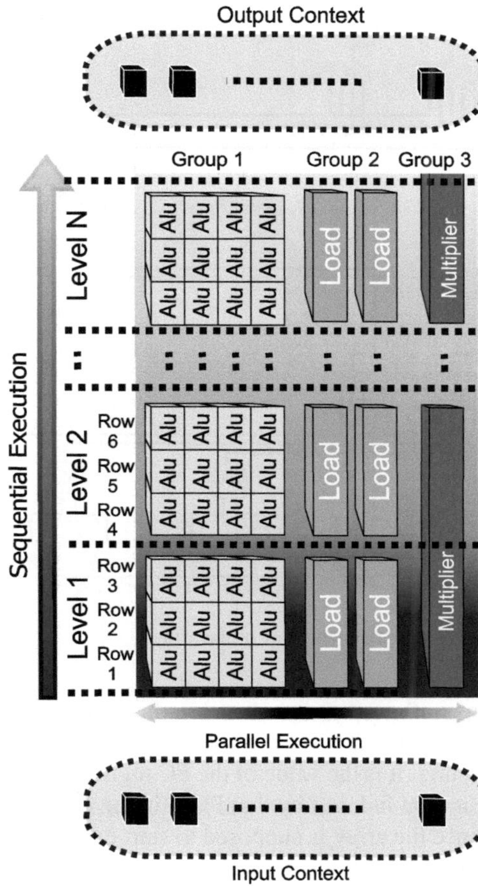

Fig. 6.2 General overview of the reconfigurable array

of which operand will be issued for execution. They are the input multiplexers. After the operation is completed, there is a multiplexer for each bus line that will choose which result will continue through that line. These are the output multiplexers. As some of the values of the input context or old results generated by previous operations can be reused by other functional units, the first input of each output multiplexer always holds the previous result of the same bus line. Note that, if one considers that the configuration of all multiplexers is set to zero at the beginning of any execution, the output context will be the same as the input context.

6.3 Translation, Reconfiguration and Execution

The reconfiguration phase involves the load of the configuration bits for the multiplexers, functional units and immediate values from the special cache, followed by the fetch of the input operands from the register bank. As already commented,

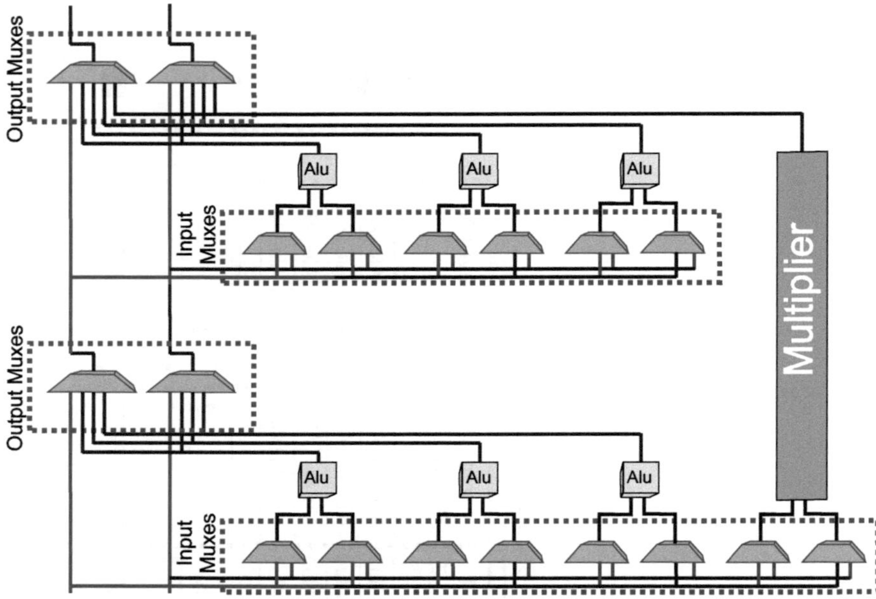

Fig. 6.3 A more detailed view of the array

a given configuration is indexed in the cache using the PC of the first instruction of the translated sequence. During execution, this address is known at the first stage of the pipeline (it is the value of the PC register). Therefore, it is possible to realize if a configuration indexed by that PC is in the configuration cache at that moment. This way, since the array is supposed to start execution in the fourth stage (the execution stage in this case), there are three cycles available for the array reconfiguration. In cases three cycles are not enough (for example, there is a large number of operands to be fetched from the register bank) the processor will be stalled and wait for the end of the reconfiguration process.

After the reconfiguration is finished, execution begins. Memory accesses are performed by the LD/ST units. Their addresses can be calculated by ALUs located in previous rows, during execution, allowing memory operations even with those addresses that are not known at compile time. The operations that depend on the result of a load are allocated considering a cache hit as the total load delay. If a miss occurs, the whole array operation is stalled until it is resolved. Finally, when the operands are not used anymore for that configuration, they are written back either in the memory or in the local registers. If there are two writes to the same register in a given configuration, just the last one will be performed, since the first one was already consumed inside the array by other instructions.

The binary translation hardware starts working on the first instruction found after a branch execution, and stops the translation when it detects an unsupported instruction or another branch (when no speculative execution is supported). If more than three instructions were found, a new entry in the cache is created and the data of a special buffer, used to keep the temporary translation, is saved. This translation relies on a set of tables, used to keep the information about the sequence of instructions that is being processed, e.g. the routing of the operands as well as the configuration of the functional units.

The BT algorithm takes advantage of the hierarchal structure of the reconfigurable array: for each incoming instruction, the first task is the verification of RAW (read after write) dependences. The source operands are compared to a bitmap of target registers of each row. If the current row and all above do not have that target register equal to one of the source operands of the current instruction, it can be allocated in that row, in a column at the first available position from the left, depending on the group. When this instruction is allocated, a dependence table is updated in the correspondent row. Summarizing the dependence information for each row, the technique increases the size of the window of instructions, which is one of the major limiting factors of ILP in superscalar processors, exactly due to the number of comparators necessary [167]. Finally, the source/target operands from/to the context bus are configured for that instruction. For each row there is also the information about what registers can be written back or saved to the memory. Hence, it is possible to write results back in parallel to the execution of other operations. Figure 6.4 shows an example of how a sequence of instructions would be allocated in the array after detection and translation. As can be observed, allocation was performed considering true data dependencies, since the BT can handle false ones.

The algorithm supports functional units with different delays and functionalities. Moreover, it also performs speculative execution. In this case, each operand that will be written back has an additional flag indicating in which speculated basic block it is located. When the branch relative to that basic block is resolved, it triggers the writes of these correspondent operands if the speculation was right. The speculative policy is based on bimodal branch predictor [179]. A saturation point is used to include and exclude basic blocks for speculation in a sequence. When the counter reaches a predefined value for a basic block candidate for speculation, the instruc-

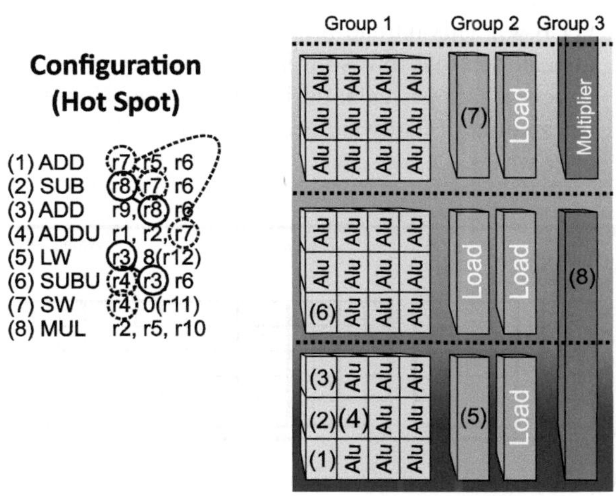

Fig. 6.4 An example of instruction allocation

tions corresponding to this basic block are added to that configuration of the array. The configuration is always indexed by the first PC address of the whole sequence. If a miss speculation happens a predefined number of times for a given basic block in a configuration, achieving the opposite value of the respective counter, that entire configuration is flushed out and the process is repeated. The BT algorithm is explained in more details in the next section.

6.4 The BT Algorithm in Details

In this subsection, the details of how the BT algorithm works are shown. To better explain it, a very basic algorithm with many restrictions, considering that the array is composed just by adders, will be first demonstrated. Then, its description will be improved until it reflects the current implementation.

6.4.1 Data Structure

The tables necessary to make the routing of the operands inside the reconfigurable array as well as the configuration of the functional units are illustrated in Fig. 6.5. In this example, it is considered an array composed of twenty ALUs (five rows with four ALUs each), although there is no restriction regarding the number of functional units available or context size. Other intermediate tables are also necessary, however, they are used only during the detection phase (their information is not saved in the cache since it is not needed during the reconfiguration phase). The tables can be described as follow:

- *Write bitmap table*: For each row, it stores information concerning data dependences. This table is in fact composed of a large number of small bitmaps (one per row). This bitmap informs what are the target registers for the instructions allocated in that row, so it is not necessary to keep this information for each instruction as superscalar designs usually do. Consequently, it is possible to reduce the amount of hardware necessary to check true data dependences (RAW—read after write).

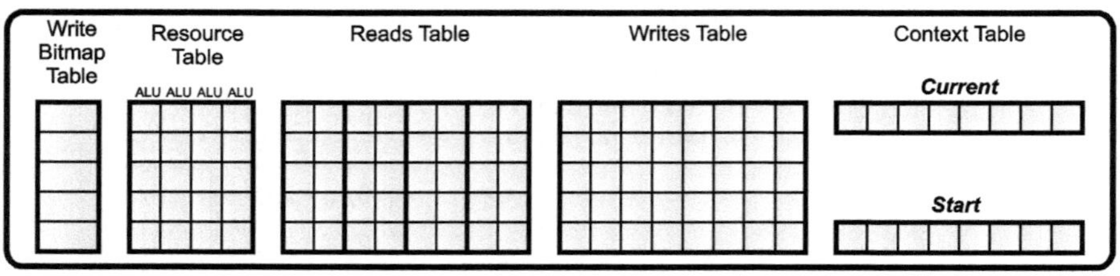

Fig. 6.5 Tables necessary for the detection and configuration of the array

- *Resource Table*: Informs if a given functional unit is free or not.
- *Reads Table*: Responsible for informing which operand from the input context must be issued to each functional unit. Considering that each functional unit has two inputs, each entry in this table holds two different values. The input context is an indirect table, meaning that not necessarily the first context slot needs to keep the value of the register R1.
- *Writes table*: This table informs what value each context slot will receive. This table is different when comparing to the previous one. In the reads table, the multiplexers were responsible for choosing which values from the context slots would be issued to each functional unit. The Writes Table, in turn, informs what values from the whole set of the functional units that compose each row will continue in each slot of the context bus.
- *Context table*: This table has only two rows. The first one represents the input context, and it will be used in the reconfiguration phase for the operands fetch. The second one is called current table, and it is used during the detection phase. Its final state represents what values will be written back when the array's execution finishes.

The tables follow the same structure as the reconfigurable array. In the case of the Resource Table, the X-axis represents the parallel execution of instructions through time (y-axis). Reads Table is almost the same, with the difference that each column of the Resource Table is split in two (as commented before, each functional unit has two input operands). Each item of the Write Table represents each slot of the Context (so the number of columns of this table is the same as the number of context slots).

6.4.2 How It Works

The following steps represent pipeline stages in the hardware implementation.

Considering that

```
inst opw, opr1, opr2
```

where inst is the current instruction and opw, opr1 and opr2 are the target and the source operands, respectively, the follow steps are necessary:

1. Decode the instruction, returning its target and source registers.
2. In the Write table, for each row from 0 to N, verify if opr1 and opr2 exist. If any one of them or both exist in the row S, row O equals to $S + 1$. Considering a bottom-up search, the row S is the last one where opr1 or opr2 appears, since they may be found in more than one row. If nor opr1 neither opr2 exist in any row of this table, row O equals to zero.
3. In the resource table, search in the columns of row O, from left to right, if there is a resource available for use. If there exists, mark that free column as C, and row R equals to O. If there is no resource available in row O, increment the value of O in 1 and repeat the same operation, until finding the free resource. This way,

row R equals to $O + N$, where N is the number of increments necessary until an available resource is found.

4. (a) Update the Write Bitmap Table in row R with the value of opw.
   (b) Update column C in row R of the resource table as busy.
   (c) Search in the current context table if there are opr1, opr2 and opw. For each one of these, if they exist, point *L1*, *L2* and W to opr1, opr2 and opw respectively. If one of them does not exist in the table, the correspondent signal of write for each one of these values in this table is set, and the correspondent pointer (*L1*, *L2* or *W*) is updated.
5. (a) Depending on the step 4c, the current context table is updated. If the pointer W is being written in the table, a flag indicating that it should be written back at the end of execution is set.
   (b) The initial context table is also updated, if one of the write signals concerning opr1 and opr2 are set.
   (c) In the Writes table, write the value of W in the row R, column C.
   (d) In the Reads table, write the values of *L1* and *L2* in row R, column C.

As a simple example, considering the following sequence of instructions:

```
add r7, r5, r6
add r8, r7, r6
add r9, r8, r6
add r1, r2, r7
add r4, r2, r7
```

The organization of the tables would be as demonstrated in Fig. 6.6, after each instruction translation. Figure 6.7 shows how this configuration would be represented in the structure of the array after its proper reconfiguration. Although the basic principles of configuration and routing remain the same, the complete version has the additional functionalities, as discussed next.

6.4.3 Additional Extensions

Immediate values are allowed for the input context. They are treated as registers in the start table. However, they cannot be changed (so, no entry for them in the current table is necessary), and hence they are fixed, being loaded together with the rest of the configuration. It is important to note that, when the upper bound limit for the input context is reached (for both immediate values and those from the register bank), a new configuration is started. Also, a new table is employed, called resource function table, with the same number of rows and columns as the resource table. This new table stores the information of which operation each functional unit will perform, so they can compute different functions.

Moreover, it is also possible to use different types of functional units, and different delays for each unit are allowed. They are divided in groups of columns, where

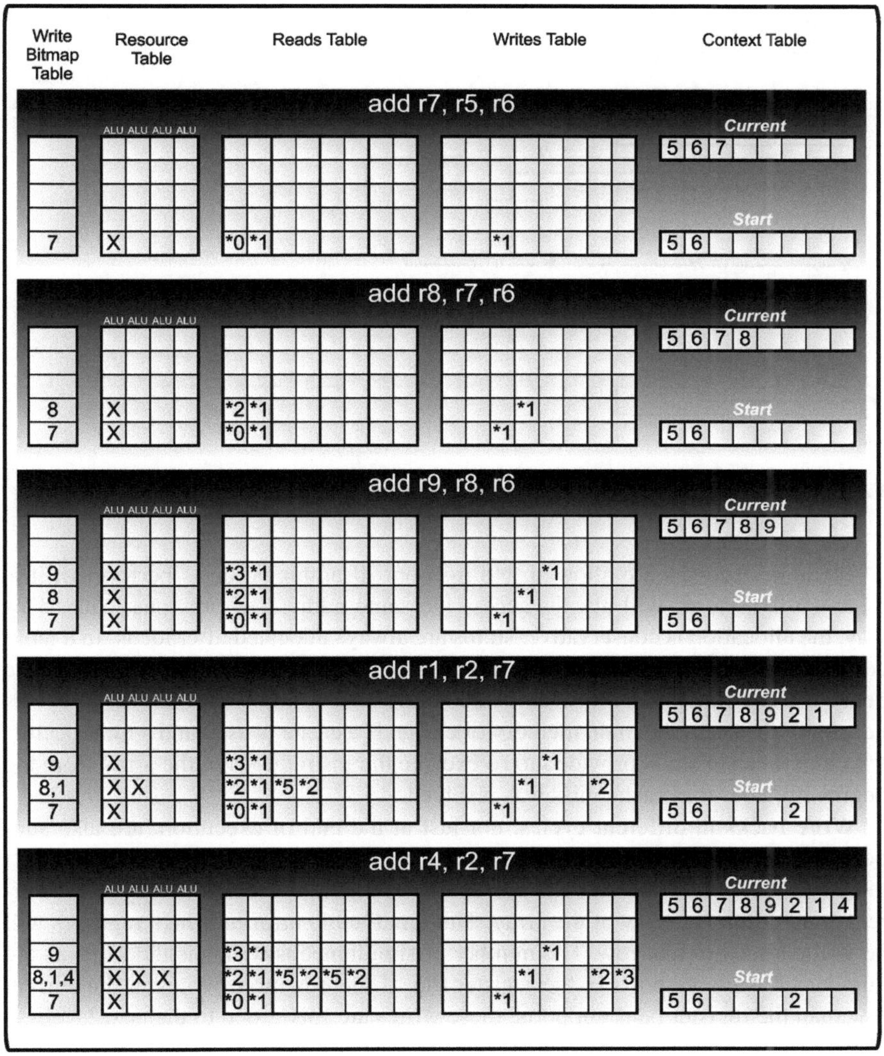

Fig. 6.6 How tables would be filled after the detection of a configuration

each column is always homogeneous, so functional units within the same group have the same delay. The resource table remains the same. However, depending on the delay a given functional unit takes for its operation, more than one row is marked as busy. For instance, if a shifter takes one processor equivalent cycle to perform its function, and the delay of three rows are equivalent to this cycle, the two positions in the same column above the one already occupied by the shift instruction will also be marked as busy.

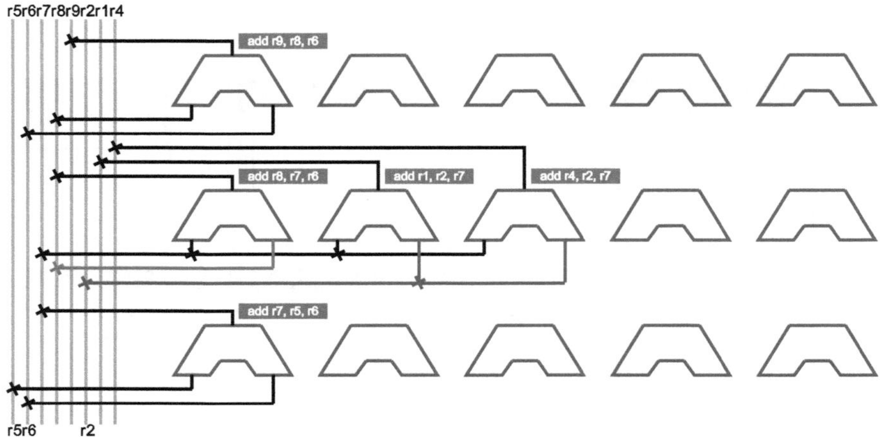

Fig. 6.7 Graphical representation of the same configuration in the array

Memory accesses are allowed. No memory disambiguation is performed: it is assumed that stores will always access the same address as previous loads. This way, the allocation is conservative: stores are always allocated after loads. In a more advanced version, however, advanced memory alias analysis will be performed. The delay of these functional units can be configured according to the number of cycles necessary to access the main memory or cache. If a cache is used and a miss occurs, a special mechanism is provided to re-execute the instructions in the array from the beginning, after that cache miss was resolved.

Write backs in different cycles, not just at the end of execution, are also supported. For that, an extension of the context table was done. Before, the context table had just two rows: start and current contexts. Now, there is a copy of the current context for each row of the array. This small table informs which registers will be written back in that row. The number of simultaneous writes is the same as the number of available ports in the register file. If there is more writes in the current row than the register bank supports, these writes are forwarded to the next level.

6.4.4 Handling False Dependencies

As already stated, the BT algorithm is capable of handling false dependencies. Let us use an example to better illustrate this approach, considering the following sequence of instructions:

```
add r7, r5, r6
sub r5, r9, r6
mul r5, r8, r6
```

Between the add and the sub, there is a false dependence, named WAR (Write After Read). In this case, the processor could not execute the sub instruction in parallel to the add because of data coherence: the value of r5 cannot be changed at the same time it is read. Between the same sub and the next instruction, mul, another type of false dependence occurs, known as WAW (Write After Write), again, with R5. Because of the same reason as before, data coherence, both instructions cannot be executed in parallel, since r5 cannot be written twice at the same time. They are declared as false dependencies because one can apply techniques to avoid them, such as Register Renaming [171]. However, register renaming is a very expensive process, and has a high cost in the design of any superscalar processor [167].

In the proposed BT algorithm, the context table is altered to easily handle false dependences. It has a pointer indicating the last operator included in the context table. When an operation needs this operator, the search occurs from the right to the left, beginning at this pointer. Each new destination operator that it is included, no matter if it is already in the table, has a new entry in the current table. In the example above, r5 would have 3 entries in the context table. If one considers that between the add and sub there would have other instructions that would read r5, they would use the first entry (included because of the add). When the sub instruction is found, a new entry of r5 is added. Any instruction between the sub and mul instruction would use this last entry in the array, because the search occurs from right to the left (from the last to the first). In the same way, any instruction executed after the mul would read the last entry of r5, and so on. As a circular buffer is employed, the previous operators that are not used anymore can be overwritten by the new ones, when necessary.

6.4.5 Speculative Execution

For speculative execution, DIM uses the same principle as trace scheduling: the configurations of the array are indexed by the PC register and the following basic blocks are speculatively executed. After execution, if there is no miss prediction, the results are written back. However, if a miss prediction occurs, meaning that a different path than the predicted before was taken, the results are discarded and the control is given back to the processor, in order to execute the instructions using its normal flow. The approach is illustrated in Fig. 6.8. In this example, it is considered that the saturation point is 2. When the Basic Block 1 (Fig. 6.8a) is found, it is allocated in the array as usual (Fig. 6.8b). After that, the branch instruction can lead to two different paths: to the Basic Block 2 or to the Basic Block 3. In this example, the path taken was to the Basic Block 3. This way, a variable responsible for that branch is incremented (equals to 1). Next time Basic Block 1 is found, the BT does not need to allocate its instructions (they have already been previously allocated). However, again, it is verified that the same branch has taken the same path as before: to the Basic Block 3. The branch control variable is incremented once more, reaching the saturation point. Consequently, the instructions of the Basic

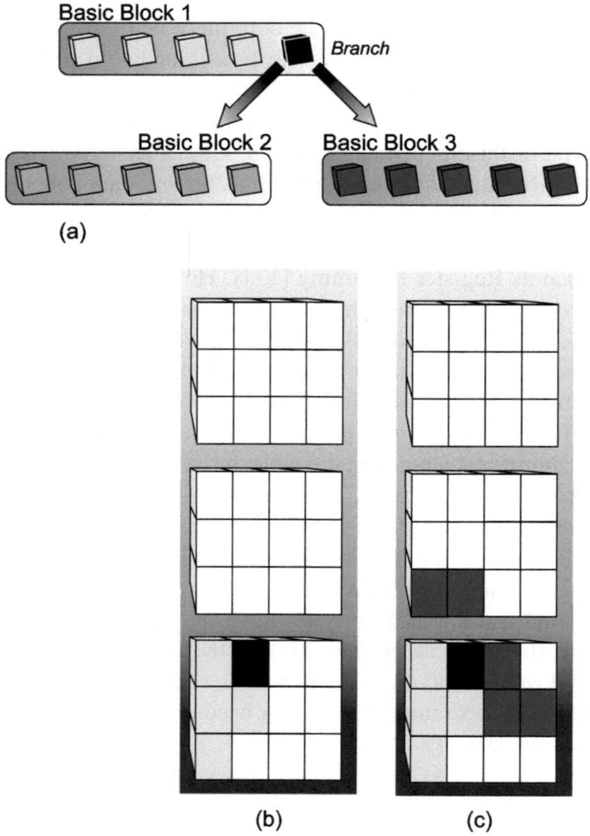

Fig. 6.8 How the saturation point works during BT detection for speculative execution

Block 3 are also allocated in the same configuration (Fig. 6.8c). On the other hand, if the path taken led to Basic Block 2, the variable would be decremented. If, during execution, the number of miss predictions in a sequence equals the saturation point for a given branch, the following basic block is removed from that configuration, starting the whole BT process again.

A new group of functional unit was created in the array, composed of branch units. For the write back of results, new multiplexers in each row were added. Without speculation, the array would have a given number of multiplexers per row directly connected to the register bank, in order to write back the results from the output context. Now, it has more multiplexers per row, divided in groups. Each group of multiplexers belongs to a given level of speculation in the array. The values of each group of multiplexers are saved in a buffer, waiting for a trigger, correspondent to the level of speculation. When a given branch unit executes the branch instruction relative to that level, the bit signal is sent, informing if the values waiting for that trigger must be written back (in case speculation was right) or should be discarded (a miss speculation happened).

This way, if there is a miss speculation, the array finishes its execution and just writes back the results of the first basic block and the ones where the result of the speculation of previous branches was correct. Then, DIM sends information to the branch controlling hardware and, instead of returning the PC of the last instruction of that configuration, it returns the one correspondent to the beginning of the basic block that was miss speculated, in order to start the execution of the non translated instructions again (now taking the right path for that branch).

6.5 Case Studies

To represent the general purpose computation field, the Simplescalar Toolset [166] was used. It simulates an out-of-order superscalar processor that executes the PISA (Portable Instruction Set Architecture) [166] which is based on the MIPS IV instruction set, so the processor is very similar to the MIPS R10000 [181]. The second architecture employed was the MIPS R3000: the classic 5-stage RISC processor, which executes the first proposed MIPS ISA. This processor is still in use, mainly in the embedded system market.

6.5.1 Coupling the Array to a Superscalar Processor

For the performance analysis, an estimator was built and integrated to the simplest version of the Simplescalar toolset: *sim-safe*. It was implemented in C (this language was used because the Simplescalar was also programmed in C), following a very similar approach that the authors in [180] used for performance estimates. Instructions are analysed at run time, during the software execution. Being totally integrated to the Simplescalar toolset, parameters concerning the reconfigurable array, such as the number of functional units and their delays, number of rows, columns, can be easily changed. In addition, the simulator provides statistics considering different reconfiguration cache sizes, giving the average time spent by several operations, such as execution, context load, write back times, etc.

In the first experiment, aimed to analyze potential speedups of the technique, the Simplescalar ToolSet was configured to behave like an ordinary in-order MIPS processor (very similar to the MIPS R3000 processor), executing a subset of the MiBench [170], with the follow algorithms: *Basicmath*, *Bitcount*, *Qsort*, *Tiffdither*, *Tiffmedian*, *Dijkstra*, *Patricia*, *Ghostscript*, *StringSearch*, *Sha*, *CRC*, *FFTinv* and *FFT*. A cache memory with three ports (two reads and one write per cycle) has been assumed, and a latency of one cycle to fetch values from the cache was also assumed. This assumption can be considered somehow very pessimistic. For instance, when presenting the trace reuse approach, the authors in [169] considered the capability to perform 16 reads and writes per cycle, including register and memory values. Superscalar architectures such as the Alpha 21264 [173] can perform up to 14 accesses per cycle (8 register reads, 4 register writes and 2 memory references).

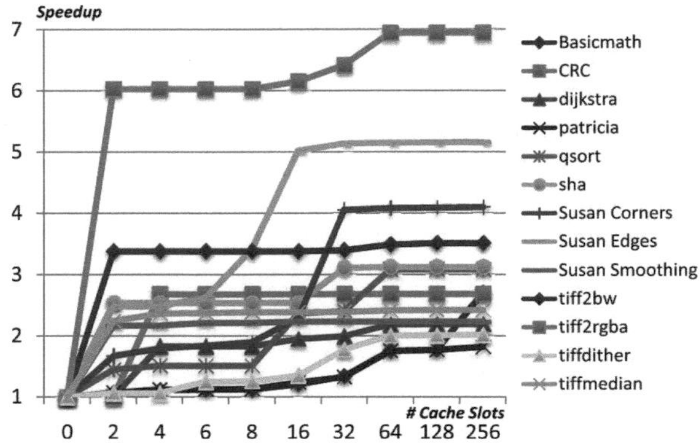

Fig. 6.9 Performance improvements when using DIM

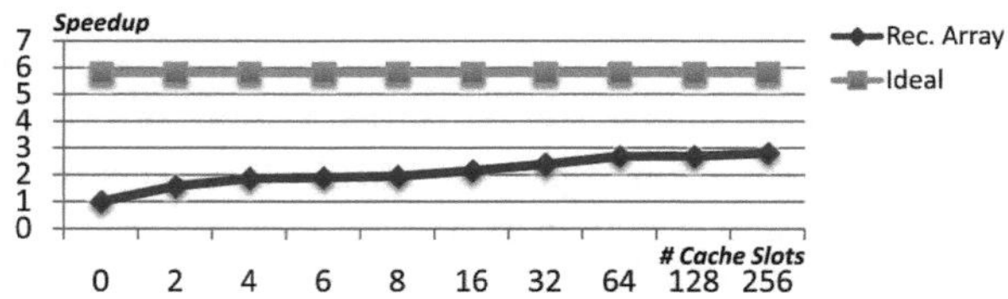

Fig. 6.10 The average performance improvements

Therefore, the employed configuration could be implemented in nowadays memory systems.

Figure 6.9 shows the performance improvements when coupling the array (with no support for speculation) to the in-order MIPS based processor, compared to its standalone version. The Y axis is the relative time spent by the algorithm according to the size of the context memory (its implementation is similar to a cache memory), shown in the X axis (zero means not using the reconfigurable array). It is considered that the array is always big enough to execute the largest configuration found. Analyzing the figure, one could notice that depending on the algorithm, a small number of context memory slots is necessary to show good performance improvements. As the context memory works as a cache, and its replacement policy implemented for this analysis was FIFO (First In, First Out), this cache must be large enough to support all the basic blocks that are being executed inside a determined period of time in order to allow their reuse. For instance, let us consider that an algorithm is composed of a main loop and inside this loop there are five basic blocks. If there are only four slots available in the context cache, the first time the first basic block will be reused (in the second iteration of the loop), it will not be in the cache anymore, and all the detection process should be done again. Therefore, in this case, no optimization would be achieved.

Figure 6.10 [161, 162] shows the average gain (Y axis) concerning all algorithms, with different context cache sizes (X axis). Depending on the context cache size, the algorithms can be executed up to three times faster. The ideal curve represents the performance gain when considering only 1 cycle per executed reconfigurable instruction.

Table 6.1 Configurations of the superscalar processor

Out of Order	
Fetch, decode and commit	up to 4 instructions
Register Update Unit	16 entries
Load/Store Queue	16 entries
Functional Units	2 Integer ALU, 1 multiplier, 2 memory ports
Branch Predictor	Bimodal/512 entries

Table 6.2 Configurations of the array

	Reconfigurable Array		
	C 1	C 2	C 3
#Lines	27	54	99
#Columns	11	16	30
#ALU / line	8	8	11
#Multiplier	1	2	3
#Ld/st / line	2	6	8

Table 6.3 IPC in the Out-of-Order processor and the average BB size

(a)

Algorithm	IPC Out-of-Order
Basicmath	1.43
CRC	2.13
dijkstra	1.76
Jpeg decode	1.86
patricia	1.40
qsort	1.79
sha	1.94
stringsearch	1.60
Susan Smoothing	1.64
Susan Corners	1.83
tiff2bw	1.90
tiff2rgba	1.92
tiffdither	1.56
tiffmedian	1.91

(b)

BB size
5.8751
7.9954
5.6011
6.2554
4.4255
4.6243
7.9381
4.8709
15.8098
13.4952
22.5567
13.4952
18.9188
30.686

The next experiment was to use the approach with support to speculative execution in the array [158, 159]. The Simplescalar was configured to behave as close as possible to the Superscalar Out-Of-Order MIPS R10000 processor, for performance comparisons. Its configuration is summarized in Table 6.1.

Table 6.2 shows three different configurations for the array employed in the experiments. For each configuration, the size of the reconfiguration cache was also changed: 2 to 512 slots, using the FIFO policy. The impact of doing speculation is evaluated considering optimization of up to two basic blocks ahead. Finally, the instruction/data cache memories were increased in order to achieve almost no cache misses, so it was possible to evaluate the results without the influence of them.

Table 6.3 shows the IPC of the out-of-order processor and the average number of branches per instructions (so one can observe the most control and dataflow oriented

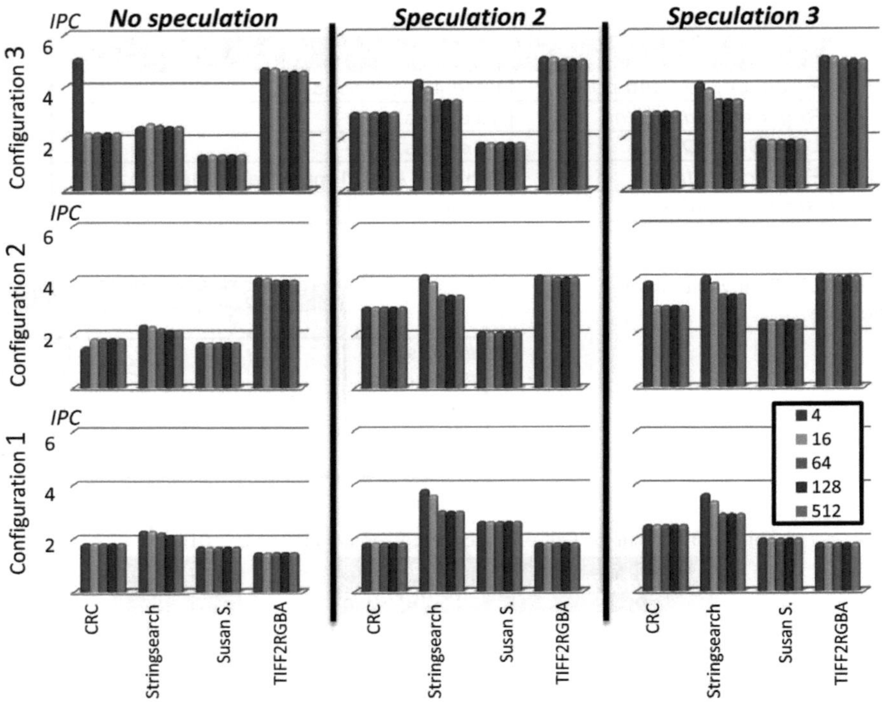

Fig. 6.11 IPC of four different benchmarks being executed in the reconfigurable logic with different configurations

algorithms). Figure 6.11 shows the IPC of the reconfigurable array, in different configurations. For each configuration, three different speculation policies are used: no speculation, 1 and 2 basic blocks ahead, varying the number of slots available in the reconfiguration cache (4, 16, 64, 128 and 512). The four benchmarks presented in this figure were chosen because they represent a very control-oriented algorithm, a dataflow one and a midterm between both, plus the CRC, which is the biggest benchmark in the subset.

As it is shown in Fig. 6.11, it is possible to achieve a higher IPC when executing instructions on the reconfigurable array in comparison to the out-of-order superscalar processor, in almost all variations. However, the overall optimization when using the proposed technique depends on how many instructions are executed on the reconfigurable logic instead of using the normal processor flow.

Table 6.4 shows the overall speedup obtained when coupling the reconfigurable array to the out-of-order processor, comparing them against the standalone out-of-order. It is important to notice that, as already discussed, reconfigurable systems usually show improvements only in very dataflow oriented programs. The DIM technique, on the other hand, can optimize both control and data oriented programs.

Table 6.4 Speedups using the reconfigurable array coupled to the out-of-order processor

Algorithm	#Cycles in the Out-Of-Order	% of Speed Up – Array with configuration 1 coupled to Out-of-Order								
		No Speculation			Speculation 2			Speculation 3		
		4	64	256	4	64	256	4	64	256
Basicmath	111169924	5.03	13.75	17.85	3.52	14.49	21.79	3.40	15.22	23.31
CRC	399531928	-16.01	-16.03	-16.03	-5.20	-5.21	-5.21	9.03	9.03	9.03
dijkstra	31094638	-22.29	-24.31	-24.33	1.30	1.25	1.25	8.45	8.46	8.46
Jpeg decode	3942226	-9.15	-9.72	-9.77	4.63	3.24	3.29	7.11	7.45	7.61
patricia	95927575	4.41	13.30	13.72	3.99	14.42	21.52	3.26	14.22	21.96
qsort	23435690	-8.76	-11.69	-11.69	-0.58	4.18	4.18	0.37	-30.41	-30.21
sha	6800950	11.56	13.07	13.07	27.22	33.45	33.45	26.30	31.29	31.29
stringsearch	115917	16.32	20.16	21.23	28.95	35.20	35.24	28.50	35.39	35.38
S. Smoothing	15628090	-0.94	-3.22	-3.22	0.31	-0.99	-1.00	2.13	1.59	1.59
S. Corners	533870	2.16	1.79	1.79	4.40	4.29	4.28	1.13	4.29	4.28
tiff2bw	27391803	-4.24	-4.38	-4.42	0.88	0.82	0.82	-0.20	-0.20	-0.20
tiff2rgba	23796384	-10.94	-11.39	-11.40	-1.53	-1.75	-1.75	-1.19	-1.39	-1.40
tiffdither	188757828	1.48	8.88	8.92	6.65	9.34	9.41	4.47	-21.46	-23.52
tiffmedian	93254386	3.95	3.74	3.73	12.91	12.82	12.82	7.42	7.38	7.38

Algorithm	#Cycles in the Out-Of-Order	% of Speed Up – Array with configuration 3 coupled to Out-of-Order								
		No Speculation			Speculation 2			Speculation 3		
		4	64	256	4	64	256	4	64	256
Basicmath	111169924	5.76	19.27	26.40	4.63	19.83	30.33	4.86	20.52	32.14
CRC	399531928	3.97	3.97	3.97	8.12	8.14	8.14	20.75	20.77	20.77
dijkstra	31094638	-21.96	-20.08	-20.04	1.00	4.34	4.36	4.13	7.65	7.67
Jpeg decode	3942226	9.76	11.92	12.05	16.55	18.94	19.06	16.77	19.51	19.68
patricia	95927575	5.06	17.97	18.89	5.25	18.80	29.07	4.57	18.58	29.80
qsort	23435690	24.29	38.95	38.95	16.79	43.74	43.74	16.44	40.72	40.72
sha	6800950	22.57	25.48	25.48	39.91	48.66	48.66	41.27	50.28	50.28
stringsearch	115917	21.02	27.05	30.57	31.25	41.02	41.17	31.04	42.61	42.63
S. Smoothing	15628090	25.35	35.66	35.69	26.87	37.95	37.96	23.73	32.05	32.04
S. Corners	533870	32.69	41.44	41.44	37.53	41.44	41.45	33.89	37.13	37.12
tiff2bw	27391803	-5.65	-5.42	-5.39	19.08	19.60	19.60	24.41	25.22	25.22
tiff2rgba	23796384	57.19	57.83	57.83	58.29	59.69	59.69	47.30	48.87	48.87
tiffdither	188757828	4.33	18.15	18.30	10.73	19.33	19.57	7.95	14.31	14.60
tiffmedian	93254386	14.13	14.11	14.13	27.23	27.43	27.43	27.36	27.72	27.72

6.5.2 Coupling the Array to the MIPS R3000 Processor

In [163], the DIM system was coupled to an improved VHDL version of the Minimips processor [165], which is based on the R3000 version. For area evaluation, the Mentor Leonardo Spectrum tool has been used, and for power estimates, Synopsis PowerCompiler, both with the TSMC 0.18u library. Data about power consumption for the main memory was taken from [174]. The system was evaluated with the Mibench Benchmark Suite [170]. All benchmarks with no representative floating point computations and that could be compiled successfully to the target architecture were utilized.

Table 6.5 shows three different array configurations used in the experiments. For each one the context cache size is varied: 16, 64 and 512 slots. The impact of performing speculation, up to three basic blocks, is also evaluated.

Table 6.6 demonstrates the speed up of the reconfigurable array for the three different configurations. It is ordered to show the most dataflow algorithms at the top and the most control flow ones at the bottom. In configuration 3 with speculation,

Table 6.5 Different configurations for the array, when coupling to the MIPS R3000

	C1	C2	C3
#rows	24	48	150
#Columns	11	16	20
#ALU / row	8	8	12
#Multipliers / row	1	2	2
#Ld/st / row	2	6	6

Table 6.6 Speedups using the reconfigurable array coupled to the MIPS R3000 processor

Algorithm	Speed Up – Configuration 1						Speed Up – Configuration 2					
	No Speculation			Speculation			No Speculation			Speculation		
	16	64	256	16	64	256	16	64	256	16	64	256
Rijndael E.	1.05	1.20	1.21	1.05	1.24	1.24	1.05	1.71	1.73	1.06	1.55	1.55
Rijndael D.	1.07	1.21	1.21	1.07	1.25	1.25	1.07	1.63	1.64	1.07	1.55	1.55
GSM E.	1.63	1.65	1.68	2.01	2.05	2.13	1.63	1.65	1.68	2.03	2.07	2.17
JPEG E.	1.95	2.04	2.07	1.79	1.88	1.89	2.50	2.72	2.77	3.55	4.27	4.37
SHA	1.90	1.90	1.90	3.81	3.84	3.84	1.90	1.91	1.91	4.80	4.84	4.84
S. Smothing	1.49	1.60	1.65	2.70	2.99	3.31	1.49	1.61	1.65	2.83	3.14	3.52
CRC	1.53	1.53	1.53	1.92	1.92	1.92	1.53	1.53	1.53	1.92	1.92	1.92
JPEG D.	1.92	2.03	2.04	1.64	1.78	1.78	2.05	2.21	2.22	2.02	2.54	2.55
Patricia	1.49	1.84	1.93	1.58	2.05	2.23	1.49	1.86	1.95	1.64	2.17	2.37
Susan Corners	1.22	1.49	1.72	1.31	1.47	1.91	1.38	1.79	2.17	1.56	1.79	2.64
Susan Edges	1.23	1.42	1.64	1.29	1.48	1.83	1.43	1.70	2.20	1.47	1.74	2.43
Dijkstra	1.59	1.71	1.71	2.03	2.21	2.22	1.59	1.72	1.72	2.04	2.24	2.24
GSM D.	1.28	1.28	1.29	1.27	1.28	1.29	1.62	1.62	1.65	1.48	1.50	1.52
Bitcount	1.76	1.76	1.76	1.83	1.83	1.83	1.76	1.76	1.76	1.83	1.83	1.83
Stringsearch	1.38	1.61	1.86	1.56	2.22	2.77	1.38	1.62	1.89	1.57	2.30	2.96
Quicksort	1.37	1.74	1.74	1.69	2.32	2.33	1.37	1.77	1.77	1.80	2.66	2.67
RawAudio E.	1.60	1.61	1.61	1.98	1.99	2.00	1.60	1.61	1.61	1.98	1.99	2.00
RawAudio D.	1.64	1.64	1.64	1.79	1.79	1.79	1.64	1.64	1.64	1.79	1.79	1.79
Average	***1.51***	***1.63***	***1.68***	***1.80***	***1.98***	***2.09***	***1.58***	***1.78***	***1.86***	***2.03***	***2.33***	***2.49***

Algorithm	Speed Up – Configuration 3						Ideal	
	No Speculation			Speculation			No Speculation	Speculation
	16	64	256	16	64	256		
Rijndael E.	1.05	3.46	3.60	1.06	2.68	2.68	5.10	8.05
Rijndael D.	1.07	3.32	3.33	1.07	2.32	2.32	4.68	7.42
GSM E.	1.63	1.65	1.69	2.03	2.07	2.19	1.70	2.19
JPEG E.	2.50	2.72	2.77	3.55	4.27	4.37	2.72	4.37
SHA	1.90	1.91	1.91	4.80	4.84	4.84	1.91	4.87
Susan Smothing	1.49	1.61	1.65	2.83	3.14	3.52	1.65	3.52
CRC	1.53	1.53	1.53	1.92	1.92	1.92	1.53	1.92
JPEG D.	2.05	2.21	2.22	2.03	2.62	2.63	2.77	4.39
Patricia	1.49	1.86	1.95	1.64	2.17	2.37	2.19	3.07
Susan Corners	1.38	1.79	2.17	1.56	1.79	2.64	2.17	2.66
Susan Edges	1.43	1.70	2.20	1.53	1.81	2.58	2.21	2.60
Dijkstra	1.59	1.72	1.72	2.04	2.24	2.24	1.72	2.25
GSM D.	2.79	2.79	2.93	2.37	2.49	2.58	3.31	3.68
Bitcount	1.76	1.76	1.76	1.83	1.83	1.83	1.76	1.83
Stringsearch	1.38	1.62	1.89	1.57	2.30	2.96	1.89	2.97
Quicksort	1.37	1.77	1.77	1.80	2.66	2.67	1.77	2.67
RawAudio E.	1.60	1.61	1.61	1.98	1.99	2.00	1.61	2.00
RawAudio D.	1.64	1.64	1.64	1.79	1.79	1.79	1.64	1.79
Average	***1.65***	***2.04***	***2.13***	***2.08***	***2.50***	***2.67***	***2.32***	***3.36***

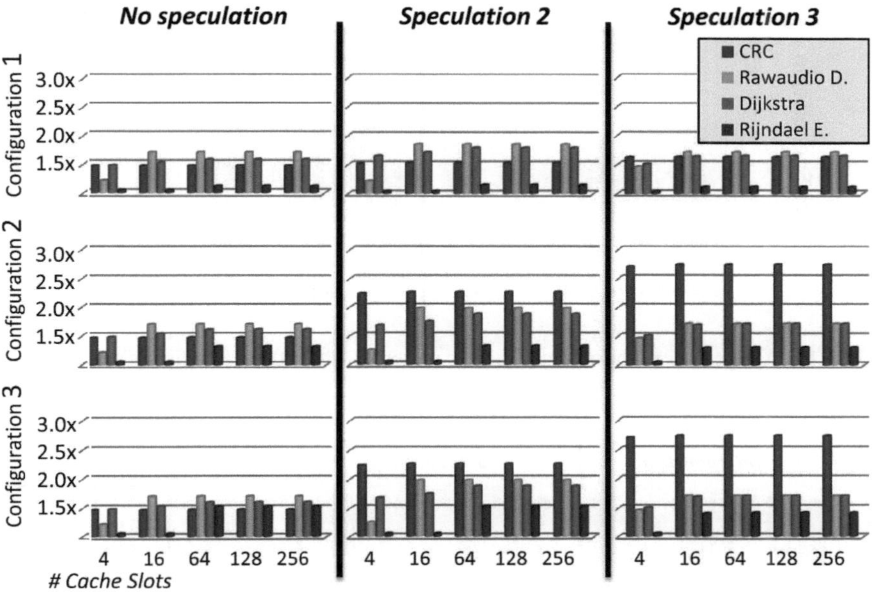

Fig. 6.12 Speedups for four different benchmarks

an average performance improvement of more than 2.5 times is achieved. Moreover, gains are shown regardless of the instruction/branch rate, even for very control oriented algorithms such as *RawAudio Decoder* and *Quicksort*, as well as those which do not have distinct kernels, such as *Susan Corners*. Together with these results, there is an extra table at the right, demonstrating the overall optimization assuming infinite hardware resources for the array. As it can be observed, with the best configuration it is possible to get very close to this theoretical speedup in several algorithms: just in five of them there is a significant difference between the most aggressive configuration and the ideal. In fact, the algorithms that can most benefit from hardware infinite resources are exactly the dataflow ones, since they demand more rows in the array, mainly when speculation is used. They have as most executed kernels basic blocks with a huge number of instructions. On the other hand, in algorithms that have no distinct kernels, the most important resource to be increased is the number of slots available in the cache memory. Figure 6.12 graphically shows the speedups for the *CRC*, *RawAudio Decoder*, *Dijkstra* and *Rijndael Decoder*, while Fig. 6.13 summarizes the results.

Figure 6.14 demonstrates the average power consumed per cycle in the Array coupled to the MIPS processor, with configurations 1 and 3 (shown as C#1 and C#3), considering 64 cache slots, and executing the algorithms *Rijndael* E., *Rawaudio D.* and *JPEG E.*, the most control and data flow ones, and a mid-term, respectively. The same Figure also shows the standalone MIPS processor, without the reconfigurable array. The consumption is shown separated for the core, data and instruction memories, reconfigurable array and cache, and BT hardware. It is interesting to note

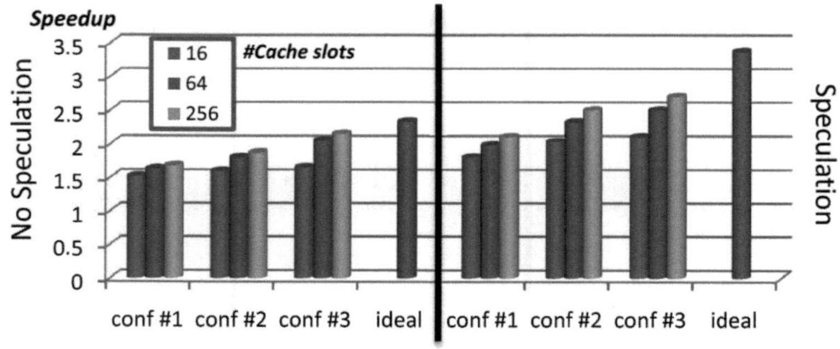

Fig. 6.13 An overview of the average speed up presented with different configurations

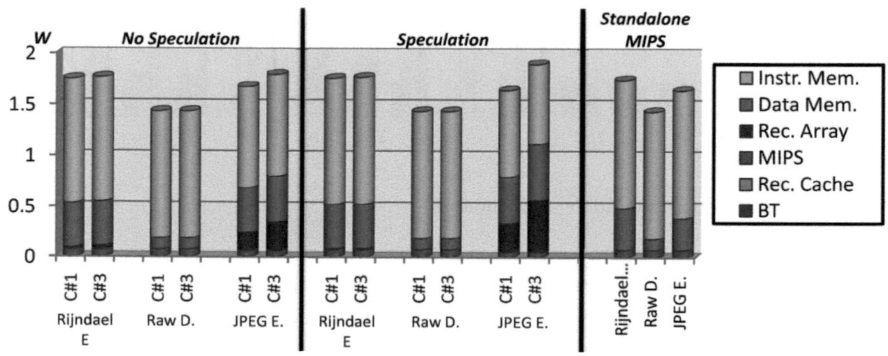

Fig. 6.14 Power consumed by 3 different algorithms in conf. 1 and 3, with and without speculation, 64 cache slots

that the major responsible for power consumption are the memory accesses. In third place comes the reconfigurable array. The power spent by this hardware depends on how much it is used during the program execution. The MIPS processor, reconfiguration cache and the BT hardware plays a minor role in this scenario.

In Fig. 6.15 the same experiment is repeated, but now analyzing the total energy consumption. As the power consumed per cycle in the array coupled to the MIPS is very similar to the standalone MIPS power consumption, but the number of cycles is reduced in the first case, energy savings are achieved. Making a deeper analysis, there are three main reasons for that:

- The execution of the instructions in a more effective way in combinational logic, instead of using the processor path.
- Avoidance of repeated parallelism analysis. As commented before, there is no necessity of performing the analysis repeatedly for the same sequence of code, since DIM saves this information in its special cache. As commented previously in this book, parallelism analysis is a huge source of power consumption.

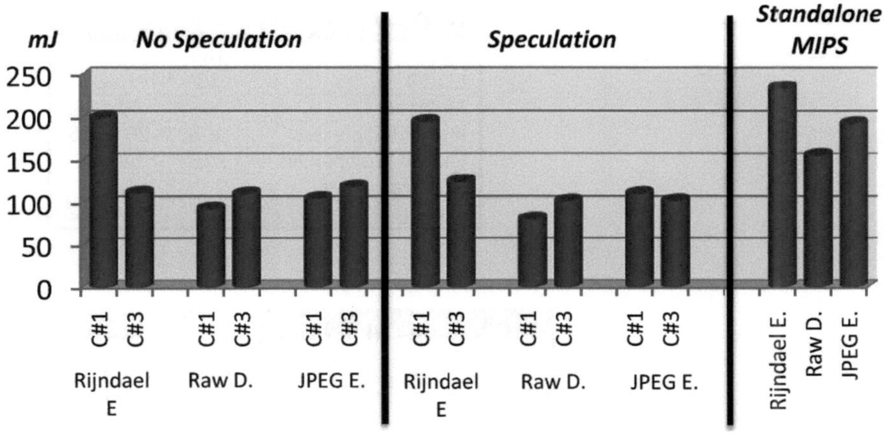

Fig. 6.15 Repeating the data of the previous figure, but now for energy consumption

- As it can be observed in Fig. 6.14, when using DIM, more power is spent in the core, because of the BT hardware, reconfigurable array and its cache. On the other hand, there are saves concerning the fetch of instructions from the memory, since they reside in the reconfiguration cache, after their proper translation to an array's configuration.

For configuration 2, with 64 cache slots, the proposed system consumes 1.73 times less energy on average than the standalone MIPS core. Moreover, assuming that the MIPS itself would be enough to handle real time constraints necessary for a given application, one could reduce the system clock frequency to achieve exactly the same performance level of the standalone processor, thus decreasing even more the power and energy consumptions.

In order to give an idea of the area overhead, Table 6.7a shows the number of functional units and multiplexers necessary to implement configuration 1, described in Table 6.5, and the number of gates they take. Table 6.7 also shows the area occupied by the DIM hardware. In Table 6.7 b the number of bits necessary to store one configuration in the context memory is presented. Note that, although 256 bits are necessary for the Write Bitmap Table, they are not added to the final total, since it is temporary and used just during detection. In Table 6.7c, the number of Bytes needed for different cache sizes is presented.

The MimiMIPS, the used version of the MIPS3000 processor used for synthesis purposes, occupies 26,712 gates. According to [181], the total number of transistors of core in the MIPS R10000 is 2.4 million. As presented in Table 6.7a, the array together with the hardware detection occupies 664,102 gates. Considering that one gate is equivalent to 4 transistors, which would be the amount necessary to implement a NAND or NOR gate, the whole system would take nearly 2.66 million transistors to be implemented.

Table 6.7 Area evaluation

Unit	#	Gates
ALU	192	300,288
LD/ST	36	1,968
Multiplier	6	40,134
Input Multiplexers	408	261,936
Output Multiplexers	216	58,752
DIM Hardware		1,024
Total		*664,102*

(a)

Table	#bits
Write Bitmap Table	256
Resource Table	786
Reads Table	1,632
Writes Table	576
Context Start	40
Context Current	40
Immediate Table	128
Total	*3,202*

(b)

#Slots	#Bytes
2	833
4	1,601
8	3,300
16	6,404
32	13,012
64	25,616
128	51,304
256	102,464

(c)

Finally, Tables 6.8 and 6.9 show, respectively, the number of functional units and the total amount of gates they would take, varying the number of rows and columns of the reconfigurable array.

6.5.3 Final Considerations

Considering all experiments, the performance numbers are mainly gathered through simulation. As these simulations are cycle accurate, the level of accuracy is high. All overheads, such as reconfiguration, context loading and write back were considered. The only exception was the simulations using *sim-safe*: in opposite to all other simulations performed, *sim-safe* works at the instruction level. This way, an average IPC was considered, based on several simulations previously performed in the cycle accurate simulator.

It is important to note that for area estimates, synthesis results from the VHDL versions of the hardware have been used, but before the place and routing phase. This way, depending on the tool methodology and technology employed, area occupation will change. Even though it is known that the system will show its full potential only when implemented as an ASIC, a prototype was implemented in the Xilinx Virtex II Pro FPGA, as a proof of concept.

Table 6.8 Number of gates according to the number of functional units

#rows	#columns			#functional units				
	ALU	LD/ST	Mult	ALU	LD/ST	Mult	MUX in	MUX out
12	8	2	1	96	8	4	192	96
	8	6	1	96	24	4	192	96
	12	6	2	144	24	8	288	96
24	8	2	1	192	16	8	384	192
	8	6	1	192	48	8	384	192
	12	6	2	288	48	16	576	192
33	8	2	1	264	22	11	528	264
	8	6	1	264	66	11	528	264
	12	6	2	396	66	22	792	264
48	8	2	1	384	32	16	768	384
	8	6	1	384	96	16	768	384
	12	6	2	576	96	32	1152	384
96	8	2	1	768	64	32	1536	768
	8	6	1	768	192	32	1536	768
	12	6	2	1152	192	64	2304	768
150	8	2	1	1200	100	50	2400	1200
	8	6	1	1200	300	50	2400	1200
	12	6	2	1800	300	100	3600	1200

#rows	#columns			#gates					
	ALU	LD/ST	Mult	ALU	LD/ST	Mult	MUX in	MUX out	*Total*
12	8	2	1	150144	2624	26756	138672	29376	***347572***
	8	6	1	150144	7872	26756	159216	29376	***373364***
	12	6	2	225216	7872	53512	225984	29376	***541960***
24	8	2	1	300288	5248	53512	277344	58752	***695144***
	8	6	1	300288	15744	53512	318432	58752	***746728***
	12	6	2	450432	15744	107024	451968	58752	***1083920***
33	8	2	1	412896	7216	73579	381348	80784	***955823***
	8	6	1	412896	21648	73579	437844	80784	***1026751***
	12	6	2	619344	21648	147158	621456	80784	***1490390***
48	8	2	1	600576	10496	107024	554688	117504	***1390288***
	8	6	1	600576	31488	107024	636864	117504	***1493456***
	12	6	2	900864	31488	214048	903936	117504	***2167840***
96	8	2	1	1201152	20992	214048	1109376	235008	***2780576***
	8	6	1	1201152	62976	214048	1273728	235008	***2986912***
	12	6	2	1801728	62976	428096	1807872	235008	***4335680***
150	8	2	1	1876800	32800	334450	1733400	367200	***4344650***
	8	6	1	1876800	98400	334450	1990200	367200	***4667050***
	12	6	2	2815200	98400	668900	2824800	367200	***6774500***

6.6 DIM in Stack Machines

The DIM technique was first proposed in [157], where both BT hardware and the reconfigurable array were coupled to the pipelined version of the Femtojava Processor [154], a processor that natively executes Java bytecodes. A large number

Table 6.9 Number of bits necessary per cache slot, varying the number of rows and columns of the array

#rows	#columns			#functional units					#bits
	ALU	LD/ST	Mult	ALU	LD/ST	Mult	MUX in	MUX out	
12	8	2	1	96	8	4	192	96	***248***
	8	6	1	96	24	4	192	96	***268***
	12	6	2	144	24	8	288	96	***357***
24	8	2	1	192	16	8	384	192	***455***
	8	6	1	192	48	8	384	192	***495***
	12	6	2	288	48	16	576	192	***672***
33	8	2	1	264	22	11	528	264	***609***
	8	6	1	264	66	11	528	264	***664***
	12	6	2	396	66	22	792	264	***908***
48	8	2	1	384	32	16	768	384	***868***
	8	6	1	384	96	16	768	384	***948***
	12	6	2	576	96	32	1152	384	***1302***
96	8	2	1	768	64	32	1536	768	***1694***
	8	6	1	768	192	32	1536	768	***1854***
	12	6	2	1152	192	64	2304	768	***2562***
150	8	2	1	1200	100	50	2400	1200	***2623***
	8	6	1	1200	300	50	2400	1200	***2873***
	12	6	2	1800	300	100	3600	1200	***3979***

of experiments have been done showing great performance improvements and reduction in energy consumption [156, 168, 176], even when compared to a VLIW version of the same architecture [155]. In addition, first studies have been done in order to optimize code at the object level using DIM [172]. Moreover, it has been shown that the DIM mechanism can take advantage of the particular computational method of stack machines in order to perform the detection with a low complexity [160]. The tool employed to provide data on the energy consumption, memory usage and performance was a configurable cycle-accurate simulator [164]. As the Java processor is a stack machine, both BT mechanism as well as the structure of the array are different from the RISC implementation.

6.7 On-Going and Future Works

The technique discussed in this chapter has been changing in several aspects to be further improved. The next sub-sections show some of the designed improvements, and demonstrate some future and ongoing work.

6.7.1 First Studies on the Ideal Shape of the Reconfigurable Array

In [177], studies about the ideal shape of the reconfigurable array considering a wide range of different applications were performed. A tool was developed to exe-

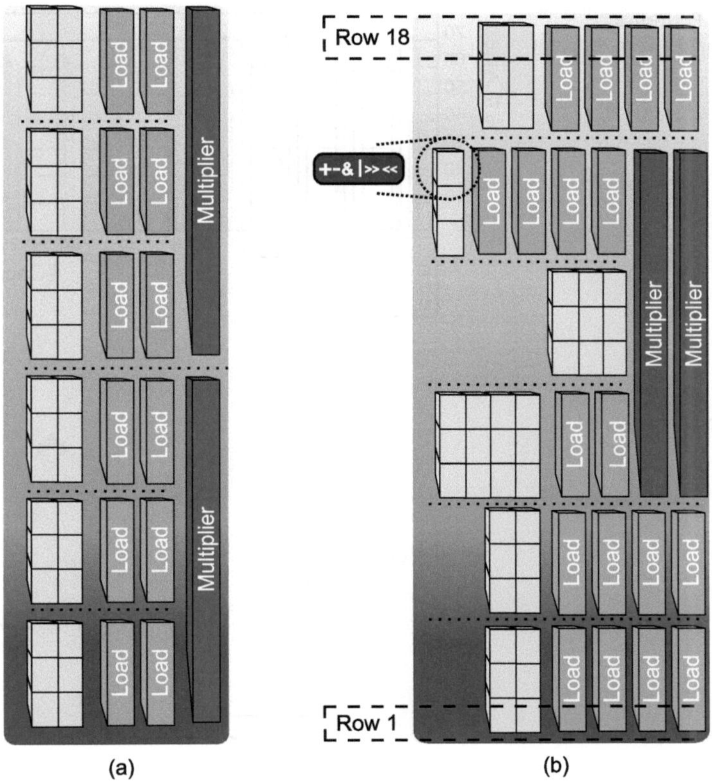

Fig. 6.16 (**a**) Original shape of the reconfigurable array (**b**) Optimized shape

cute several algorithms with different behaviors in order to find the best placement of functional units inside the array, with the minimal performance loss possible. Significant results were achieved concerning the area occupied by the system. Figure 6.16a shows the original shape of the array (rectangular). Figure 6.16b demonstrates it after the optimization analysis.

Figure 6.17 shows the average gain in performance considering all the benchmarks. Three different shapes were considered, varying the cache memory size. The gains are relative to the standalone MIPS processor, without the array. It is possible to observe that the new shape has a small performance loss (5.8% on average) when comparing to the reconfigurable array 1 (original rectangular shape with a large amount of functional units). However, the new shape has a performance improvement of 3% over the reconfigurable array 2, which is also based on the rectangular shape, but with a similar number of functional units as the new shape has. Even though this relative gain appears to be low, it is important to point out that there is an area reduction of almost 15%. In other words, when considering a similar number of functional units, the new shape presents a small performance improvement and a considerable area reduction when comparing to its original form.

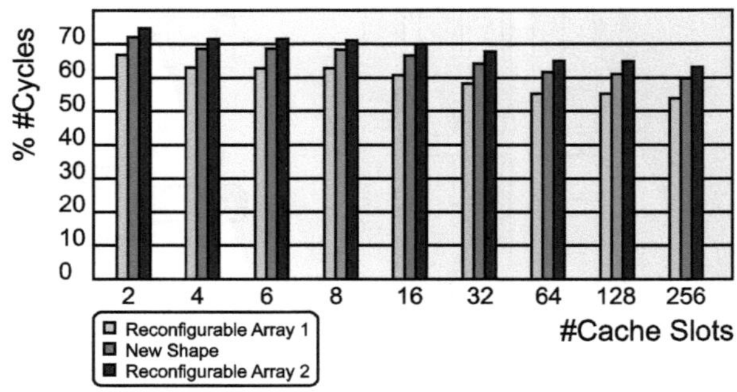

Fig. 6.17 Performance comparison between different shapes

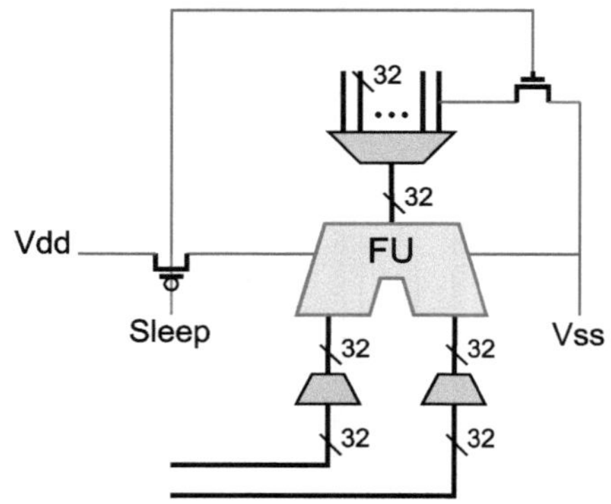

Fig. 6.18 Sleep Transistor coupled to a Functional Unit

6.7.2 Sleep Transistors

In [175] the DIM approach was used together with sleep transistors [178], so it would be possible to decrease the power consumed by inactive functional units. The BT algorithm was changed so it is aware of the existence of sleep transistors in the array. Therefore, both leakage and dynamic power consumption are avoided, since the power supply (Vdd) will not be present as well as there will be no switching activity in their inputs. In the proposed system, each functional unit can be switched off individually. Figure 6.18 shows its schematic block diagram. The area overhead to build this low power architecture is of only two transistors per functional unit.

To use such approach, only a small modification in the BT algorithm has been necessary. An extra bit was added for each functional unit. These bits correspond to each functional unit status (on or off). At the beginning of execution, all functional units are turned off. When the BT algorithm allocates an instruction in a functional unit of the array, the bit for the sleep transistor that corresponds to that FU is set. Therefore, when a given configuration is executed on the reconfigurable architecture, all the functional units that have not been allocated to an instruction are kept in the off state by the sleep transistors.

6.7.3 Speculation of Variable Length

The speculation performed in the reconfigurable array has a fixed length, meaning that it will always speculatively execute a fixed number of basic blocks ahead, even if there are still free resources available in the array. However, with small modifications in the BT algorithm, it is possible to make the speculation with a variable length concerning the basic blocks. For instance, the BT could speculate until there are no more resources available in the array. This change would also influence the reconfiguration cache.

6.7.4 DSP, SIMD and Other Extensions

Following a trend that can be observed in some of the reconfigurable systems found in these days, this topic relates to the study about adding DSP and others extensions in the array. This can be achieved by adding new hardware or optimizing the routing mechanism (for instance, in the case of Multiply-and-Accumulate operations). The BT hardware should be adapted in order to detect these new instructions. In the case of SIMD instructions, one alternative would be to add a new group of functional units supporting a larger word length. This approach could be extended to any kind of special instruction depending on the desired field of application. The greatest advantage of such extensions is that the array does not lose its backward compatibility: if the new extension is not used, the only drawback is the unused extra area overhead.

6.7.5 Design Space to Be Explored

There is still a huge design space to be explored, with a lot of open questions, such as:

- What is the best tradeoff considering the amount of instructions that can be executed in parallel, or how many write backs to the registers or memory per cycle is the ideal;
- How many configurations the context cache can store, and what is the best replacement policy for it;
- How deep is the ideal speculation regarding basic blocks;
- As all these configurations can be different depending on the characteristics of a given benchmark, other benchmark sets should be evaluated, such as SPEC.

References

154. Beck, A.C.S., Cairo, L.: Low power java processor for embedded applications. In: VLSI-SOC: From Systems to Chips. IFIP International Federation for Information Process-

ing, vol. 200, pp. 213–228. Springer, New York (2006). http://www.springerlink.com/content/14rh612330184tu8/
155. Beck, A.C.S., Carro, L.: A vliw low power java processor for embedded applications. In: SBCCI'04: Proceedings of the 17th Symposium on Integrated Circuits and System Design, pp. 157–162. ACM, New York (2004). doi:10.1145/1016568.1016614
156. Beck, A.C.S., Carro, L.: Application of binary translation to java reconfigurable architectures. In: IPDPS'05: Proceedings of the 19th IEEE International Parallel and Distributed Processing Symposium (IPDPS'05)—Workshop 3, p. 156.2. IEEE Computer Society, Los Alamitos (2005). doi:10.1109/IPDPS.2005.111
157. Beck, A.C.S., Carro, L.: Dynamic reconfiguration with binary translation: breaking the ilp barrier with software compatibility. In: DAC'05: Proceedings of the 42nd Annual Design Automation Conference, pp. 732–737. ACM, New York (2005). doi:10.1145/1065579.1065771
158. Beck, A.C.S., Carro, L.: Transparent acceleration of data dependent instructions for general purpose processors. In: IFIP VLSI-SoC 2007, IFIP WG 10.5 International Conference on Very Large Scale Integration of System-on-Chip, Atlanta, GA, USA, 15–17 October 2007, pp. 66–71. IEEE Press, New York (2007)
159. Beck, A.C.S., Carro, L.: Reconfigurable acceleration with binary compatibility for general purpose processors. In: VLSI-SoC: Advanced Topics on Systems on a Chip. IFIP International Federation for Information Processing, vol. 291, pp. 1–16. Springer, New York (2009). http://www.springerlink.com/content/p17618617681uvx3/
160. Beck, A.C.S., Gomes, V.F., Carro, L.: Exploiting java through binary translation for low power embedded reconfigurable systems. In: SBCCI'05: Proceedings of the 18th Annual Symposium on Integrated Circuits and System Design, pp. 92–97. ACM, New York (2005). doi:10.1145/1081081.1081109
161. Beck, A.C.S., Gomes, V.F., Carro, L.: Automatic dataflow execution with reconfiguration and dynamic instruction merging. In: IFIP VLSI-SoC 2006, IFIP WG 10.5 International Conference on Very Large Scale Integration of System-on-Chip, Nice, France, 16–18 October 2006, pp. 30–35. IEEE Press, New York (2006)
162. Beck, A.C.S., Gomes, V.F., Carro, L.: Dynamic instruction merging and a reconfigurable array: Dataflow execution with software compatibility. In: Reconfigurable Computing: Architectures and Applications. Lecture Notes in Computer Science, vol. 3985, pp. 449–454. Springer, Berlin/Heidelberg (2006). http://www.springerlink.com/content/86458544617q0366/
163. Beck, A.C.S., Rutzig, M.B., Gaydadjiev, G., Carro, L.: Transparent reconfigurable acceleration for heterogeneous embedded applications. In: DATE'08: Proceedings of the Conference on Design, Automation and Test in Europe, pp. 1208–1213. ACM, New York (2008). doi:10.1145/1403375.1403669
164. Beck Fl., A.C.S., Mattos, J.C.B., Wagner, F.R., Carro, L.: Caco-ps: A general purpose cycle-accurate configurable power simulator. In: SBCCI'03: Proceedings of the 16th Symposium on Integrated Circuits and Systems Design, p. 349. IEEE Computer Society, Los Alamitos (2003)
165. Bem, E.Z., Petelczyc, L.: Minimips: a simulation project for the computer architecture laboratory. In: SIGCSE'03: Proceedings of the 34th SIGCSE Technical Symposium on Computer Science Education, pp. 64–68. ACM, New York (2003). doi:10.1145/611892.611934
166. Burger, D., Austin, T.M.: The simplescalar tool set, version 2.0. SIGARCH. Comput. Archit. News **25**(3), 13–25 (1997). doi:10.1145/268806.268810
167. Burns, J., Gaudiot, J.L.: Smt layout overhead and scalability. IEEE Trans. Parallel Distrib. Syst. **13**(2), 142–155 (2002). doi:10.1109/71.983942
168. Gomes, V.F., Beck, A.C.S., Carro, L.: Trading time and space on low power embedded architectures with dynamic instruction merging. J. Low Power Electron. **1**(3), 249–258 (2005)
169. Gonzalez, A., Tubella, J., Molina, C.: Trace-level reuse. In: ICPP'99: Proceedings of the 1999 International Conference on Parallel Processing, p. 30. IEEE Computer Society, Los Alamitos (1999)

170. Guthaus, M.R., Ringenberg, J.S., Ernst, D., Austin, T.M., Mudge, T., Brown, R.B.: Mibench: A free, commercially representative embedded benchmark suite. In: Workload Characterization, 2001. WWC-4. 2001 IEEE International Workshop on, pp. 3–14 (2001)
171. Hennessy, J.L., Patterson, D.A.: Computer Architecture, 4th edn. A Quantitative Approach. Morgan Kaufmann, San Mateo (2006)
172. de Mattos, J.C.B., Beck, A.C.S., Carro, L.: Object-oriented reconfiguration. In: 18th IEEE International Workshop on Rapid System Prototyping (RSP 2007), 28–30 May 2007, Porto Alegre, RS, Brazil, pp. 69–74. IEEE Computer Society, Los Alamitos (2007)
173. McLellan, E.J., Webb, D.A.: The alpha 21264 microprocessor architecture. In: ICCD'98: Proceedings of the International Conference on Computer Design, p. 90. IEEE Computer Society, Los Alamitos (1998)
174. Puttaswamy, K., Choi, K.W., Park, J.C., Mooney III, V.J., Chatterjee, A., Ellervee, P.: System level power-performance trade-offs in embedded systems using voltage and frequency scaling of off-chip buses and memory. In: ISSS'02: Proceedings of the 15th International Symposium on System Synthesis, pp. 225–230. ACM, New York (2002). doi:10.1145/581199.581249
175. Rutzig, M.B., Beck, A.C., Carro, L.: Dynamically adapted low power asips. In: ARC'09: Proceedings of the 5th International Workshop on Reconfigurable Computing: Architectures, Tools and Applications, pp. 110–122. Springer, Berlin/Heidelberg (2009)
176. Rutzig, M.B., Beck, A.C.S., Carro, L.: Transparent dataflow execution for embedded applications. In: ISVLSI'07: Proceedings of the IEEE Computer Society Annual Symposium on VLSI, pp. 47–54. IEEE Computer Society, Los Alamitos (2007). doi:10.1109/ISVLSI.2007.98
177. Rutzig, M.B., Beck, A.C.S., Carro, L.: Balancing reconfigurable data path resources according to application requirements. In: 22nd IEEE International Symposium on Parallel and Distributed Processing, IPDPS 2008, Miami, Florida, USA, April 14–18, 2008, pp. 1–8. IEEE Press, New York (2008)
178. Shi, K., Howard, D.: Challenges in sleep transistor design and implementation in low-power designs. In: DAC'06: Proceedings of the 43rd Annual Design Automation Conference, pp. 113–116. ACM, New York (2006). doi:10.1145/1146909.1146943
179. Smith, J.E.: A study of branch prediction strategies. In: ISCA'98: 25 Years of the International Symposia on Computer Architecture (Selected Papers), pp. 202–215. ACM, New York (1998). doi:10.1145/285930.285980
180. Tiwari, V., Malik, S., Wolfe, A.: Power analysis of embedded software: a first step towards software power minimization. Readings in hardware/software co-design, pp. 222–230 (2002)
181. Yeager, K.C.: The mips r10000 superscalar microprocessor. IEEE Micro **16**(2), 28–40 (1996). doi:10.1109/40.491460

Chapter 7
Conclusions and Future Trends

Abstract Besides concluding the book, this final chapter discusses different ideas and new trends of reconfigurable architectures, such as the impact of new routing mechanisms, how reconfigurable computing will eventually merge with multi processors architectures, and how they will be used in the near future when the connection with future unreliable and non-scalable technologies must be done.

7.1 Introduction

This book presented several techniques that are candidates to be employed in a near future. First, challenges and main motivations to use reconfigurable devices were discussed. Then, the principles of reconfigurable systems, their potential and classification were presented. After that, a large number of reconfigurable systems was demonstrated. However, it has been shown that to reach a widespread use, such architectures must somehow adapt to the applications, and even to changing patterns of the same application during its execution. Hence, dynamic techniques became necessary. That was the reason that binary translation and reuse (of instructions, basic block, or traces) were discussed. Finally, architectures that jointly use both ideas of reconfiguration and dynamic optimizations were shown, including detailed analysis of one: the DIM architecture. In this chapter, some future directions and new research topics about the techniques presented before are discussed.

7.2 Decreasing the Routing Area of Reconfigurable Systems

It has been shown that for the commercially available FPGAs, routing is a very important factor for what concerns area and power consumption [183]. Nevertheless, the routing impact can also be observed in coarse grain architectures. For instance, depending on the configuration used in [184], the multiplexers are responsible for almost half the total area of the reconfigurable system.

A.C. Schneider Beck Fl., L. Carro, *Dynamic Reconfigurable Architectures and Transparent Optimization Techniques*,
DOI 10.1007/978-90-481-3913-2_7, © Springer Science+Business Media B.V. 2010

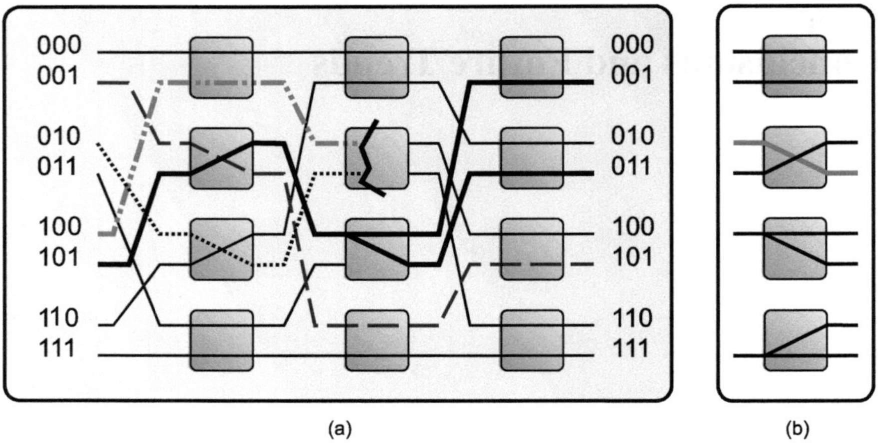

Fig. 7.1 (**a**) 8 × 8 Omega Network; (**b**) Four Switch States

This way, different structures have been proposed in order to decrease the impact of routing resources. For example, in [189, 190] the employment of Multistage Interconnection Network (MIN) was proposed to be used at the word level, on a coarse-grained reconfigurable architecture. MINs have been successfully used in several computer system levels and applications in the past. The approach takes into account one-to-one as well as multicast (one-to-many) permutations, and can handle blocking and non-blocking networks, symmetrical or asymmetrical topologies. Besides proposing the use of MIN at the register/ALU level, a new parallel self-placement and routing mechanism that runs in real time during reconfiguration execution that can be used in any kind of MIN was also presented.

The general overview of the implemented MIN, an $N = 2^k$ input/output Omega Network, is shown in Fig. 7.1a. Each internal switch can assume four states, as demonstrated in Fig. 7.1b. Considering that the network consists of $\lg_2 N$ stages of $N/2$ switches, where i is the row number and j is the column or stage level, the stages are interconnected by the following pattern: *switch output* (i, j) is connected to the *switch input* $(2*i, j+1)$ when $0 < i < N/2$; and to *switch input* $((2*i)N+1, j+1)$ when $i > N/2$. This way, the total number of switches is $N/2 \lg_2 N$. In the illustrated example, input 1 (001) is connected to output 5 (101), and input 5 (101) is connected to outputs 1 and 3. As in this example it is shown an one-to-many MIN, one input can send data to one or more outputs. As can also be observed, there is a collision at line 3, stage 1, when input 2 (010) tries to connect to output 7, and input 4 (100) tries to connect to output 6 (101). Therefore, by adding an extra layer the hardware responsible for the routing can find a path, with minimal hardware overhead.

Replacing the multiplexers (that compose a structure very similar to a crossbar) for the MIN in the architecture proposed in [184], an overall area reduction of 30% without incurring in any performance overhead was presented. Albeit the case study was applied to a specific reconfigurable system, it could be easily extended to be employed in any coarse grain architecture, such as Morphosys [194], Piperench [191], and other two dimensional structures.

Another trend is the use of LUTs with more inputs. This way, there will be more computation within one single computational block, decreasing the amount of communication necessary between them. For instance, in [182] a study on the number of LUT inputs and cluster size have been performed.

7.3 Measuring the Impact of the OS in Reconfigurable Systems

Operating Systems (OS) have been used in general purpose computing for decades. On the other hand, the simple OS employed in embedded systems is being replaced for more complex ones. That happens because embedded devices with multiple functionalities are becoming a market mainstream. Multimedia applications, communication protocols, input/output connectivity, all applications executing on a portable device, possibly at the same time, exemplify the complex control needed to manage these devices. Because of that, operating systems are being used as a fast design solution to solve the difficult task of managing systems with several different resources. Nevertheless, as OS is an extra software layer between the hardware and the final application, performance, power and memory footprint are certainly stressed.

Reconfigurable architectures should also be able to optimize OS code, such as system calls. However, there is the problem of source code availability: usually, traditional reconfigurable systems need the source code, so that they can transfer and later optimize the code through execution in reconfigurable logic. Nevertheless, some of the most used OS in the market do not have their source code available, or recompiling it could be a huge task. Considering these motivations, in [184], the two following questions are addressed:

- How much impact causes the OS routines in the execution time of the traditional embedded applications?
- For the sake of hardware efficiency, how could the same reconfigurable hardware be used to accelerate both embedded applications as well as OS routines?

As case study, the reconfigurable system used in [184] was coupled to a MIPS R4000 processor running an embedded Linux distribution. Figure 7.2 shows the applications speedup considering several point of views. The leftmost bar illustrates the speedup obtained using the accelerator when considering only instructions executed in the user mode. The second bar demonstrates the OS (Linux) code optimization. In four applications the Linux optimization takes more cycles than the user code, since many services are requested. If *bitcount* presents a higher speedup for the user than the kernel code, *stringsearch* shows lower speedups when one considers only the user code: a lot of time is spent in the kernel code. The average speedup for Linux code is 2.70 times, while for user code is 2.84 times.

The third bar in Fig. 7.2 shows the speedup factor considering the optimization of both user and kernel codes. This bar is strongly related to the dominant executed

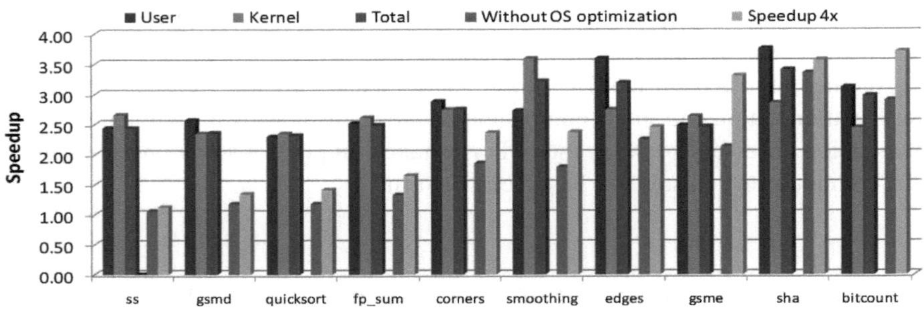

Fig. 7.2 How representative the OS routines are

code of each application. For instance, the total speedup bar of *bitcount* tends to the leftmost bar, since, in this case, it represents the dominant executed code of the application (user mode). On average, all applications present 2.76 times of performance improvements.

To stress the importance of OS optimization, the fourth bar demonstrates the environment found in traditional reconfigurable approaches. In this bar, only the user code of the applications are accelerated, while OS code is executed as normal instructions in the regular processor flow. On average, the reconfigurable system would present performance boosts of only 1.5 times. To reinforce the same idea, the last bar of each algorithm, in the same figure, simulates a four times speedup factor in the user code applications, still without any OS code acceleration. This bar aims to indicate that even high speedups achieved by a reconfigurable system, but limited to the user code, produce poor overall acceleration. For instance, the four times speedup factor for user code in *stringsearch*, *gsmd*, *quicksort* and *fpsum* would bring a speedup of only 1.38 times considering total code execution.

7.4 Reconfigurable Systems to Increase the Yield

One of the major problems that industry faces nowadays is the decrease in the yield rate. The cost of processors is directly influenced by the faults that occur during the manufacturing process. Considering GPPs, if a single fault occurs in its control or datapath, the whole processor must be discarded. On the other hand, if the fault happens in the cache memory and the processor has, for instance, two separate banks, it still may be used.

The device's miniaturization also increases the fault rates. The scaling process shrinks the wire's diameter. Besides making them more fragile and susceptible to break, it is also harder to keep contact integrity between wires and devices. According to Borkar [185] in a 100 billion transistor device, 20 billion will fail in the manufacture while 10 billion will fail in the first year of operation. The authors in [188] state that at nano-scale basis, the defect rate should be around 1% to 15% for wires and connections.

Therefore, several approaches replicate hardware in order to maintain proper circuit working. However, traditional redundancy techniques based on resources replication, such as N-Modular Redundancy, can be extremely costly, not only because of the large amount of area required to tolerate high defect densities [187], but also for the excessive power dissipation they will bring.

Reconfigurable architectures are strong candidates to cope with these problems. First, they consist essentially of identical functional elements. This regularity can be exploited as spare-parts, as it has been done in memory devices for a long time now [196]. Moreover, the reconfiguration capability can be exploited to change the resources allocation based on the position of the defective elements. At the same time, since reconfigurable architectures can adapt their behavior according to the application, this characteristic can be exploited to amortize the performance degradation caused by the replacement of defective resources.

Several and different solutions could be applied to be used with such systems. For instance, a test in the reconfigurable fabric could be performed at the beginning of execution, using an algorithm to test all parts of the reconfigurable array, marking the non-functional ones. As the array occupies the majority of the die area, it is very likely that any fail will occur at the reconfigurable part, thus increasing the overall yield rate.

Up to now, the efforts are concentrated in tolerating permanent fabrication defects. The approach consists in avoiding the defective functional units and interconnection elements and replacing them by operational ones at the cost of a performance penalty caused by the reduction of available resources. Several works propose the use of run-time reconfiguration to fault tolerance, and most of the them are applied to fine grain FPGAs [186]. In [192] a mechanism to replace defective elements in a dynamic system with a coarse grain array is proposed, with almost no performance overhead introduced by the defect tolerance approach and reduction of available resources. A performance analysis demonstrated that under a 20% defect rate, the reconfigurable system was capable of sustaining the same performance. The main idea of the approach is based on marking defective units as being always busy, so no operations will be allocated in them.

7.5 Study of the Area Overhead with Technology Scaling and Future Technologies

This study concerns the analysis of the area overhead of reconfigurable architectures according to future technologies. What is the impact of using the reconfigurable systems with an even larger area available in a near future? Furthermore, what are the possibilities of implementing them using other technologies instead of silicon? Considering the fact that the array is very regular and easily scalable, could that be an advantage?

7.6 Scheduling Targeting to Low-power

When building a reconfigurable instruction, the scheduling is done in a way to achieve the highest possible level of parallelism. However, instead of trying to reach the maximum performance, the scheduler could try to place instructions in the reconfigurable logic with the objective of keeping the largest possible number of basic reconfigurable units turned off—decreasing the power consumed by the system. For example, let us consider a coarse grain system. In a given configuration, there is an opportunity of executing two instructions in parallel, but one of the functional units necessary for that operation is turned off (the previous configuration was not using it). This way, instead of allocating the instruction in that functional unit, another one would be chosen, probably taking slightly more time to execute the current configuration, but saving power.

7.7 Granularity—Comparisons

Is a coarse grain reconfigurable system faster than a FPGA based one? If one considers that the granularity of the first is coarser than the second, a simple operation would be executed faster in a coarse grain array. However, at bit manipulation, FPGAs tend to obtain an advantage. Another issue is the routing. As already discussed, FPGAs tend to spend a lot of routing resources. Moreover, what would be the differences when, using some kind of generic tool (a fair environment) to build the very same configuration for both fine and coarse grain arrays, and executing diverse types of algorithms, that work at bit and word levels?

Another particular issue is to compare static FPGA systems against dynamic coarse ones. FPGA synthesis tools are more intelligent and have more time to build a configuration. This would be an advantage that could overcome some of the routing and allocation problems cited before. A coarse grain and dynamic system, on the other hand, needs to use a fixed structure and does not have any time to optimize it, hence the routing algorithm should be as simple as possible.

7.8 Reconfigurable Systems Attacking Different Levels of Instruction Granularity

7.8.1 Multithreading

The search for processing power in a limited design space has also been modifying the whole paradigm of parallelism exploitation. The parallelism grain is not explored just at the instruction level anymore, but also at threads and processes levels. To better illustrate what would be the difference between a regular reconfigurable architecture and systems based on the multithreaded and simultaneous multithreading

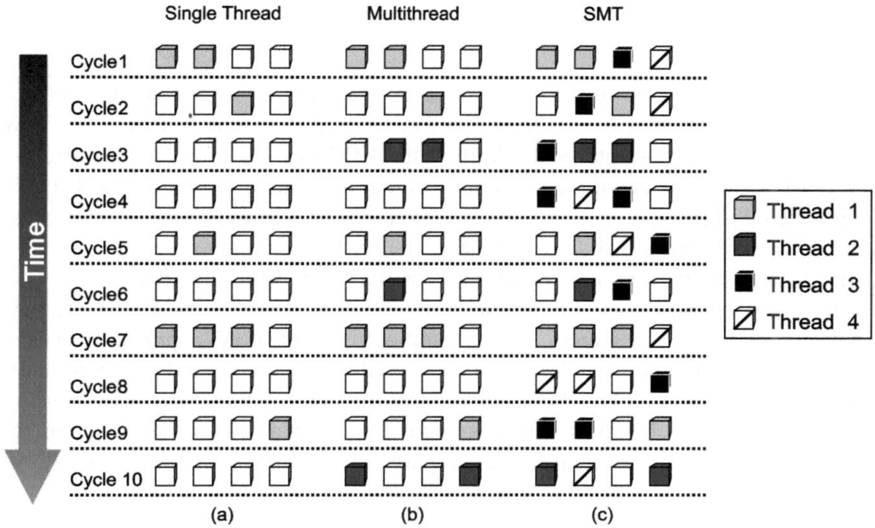

Fig. 7.3 Configurations for different models and their functional units executing various threads

(SMT) approaches, Fig. 7.3 demonstrates three different configurations considering a coarse grain reconfigurable system. Each square represents one functional unit of the reconfigurable unit. If one square is filled, it means that the correspondent functional unit was used. When it is empty, that functional unit was idle at that time.

Bringing the concept presented in [197] to the reconfigurable field, non-used functional units can be characterized as horizontal or vertical waste. Horizontal waste occurs when one or more functional units are not used within a row. Vertical waste means that all units within a given row are not used at all (the whole row was wasted). Figure 7.3a shows a configuration of a regular reconfigurable system that executes only one thread at a time. In this configuration, horizontal waste can be observed in rows 1, 2, 5, 7 and 9, while the vertical waste can be seen in rows 3, 4, 6, 8 and 10.

In Figs. 7.3b and 7.3c the configurations of the multithread and SMT processors are shown, respectively. A multithread reconfigurable architecture could be able to allocate instructions from different threads in different rows, helping to avoid vertical waste. A SMT reconfigurable architecture, on the other hand, could allocate instructions from different threads within the same rows. This way, if there is a limit in the ILP that can be explored in one thread, the functional units can be fed from others. By consequence, the horizontal waste is also dramatically reduced. Examples of SMT implementations in general purpose computation are: Intel Pentium 4 and Core i7 (which technology is called Hyperthreading), Alpha EV8, IBM Power 5 and Sun Microsystems' UltraSPARC T1. In [195], the authors started the study on this subject considering reconfigurable systems, extending the Warp processing technique to support several different threads to be executed concurrently.

7.8.2 CMP

As can also be observed in these days, superscalar processors are giving space to CMP (Chip Multi Processing), sometimes composed of simpler processors. New processors produced by Intel and AMD, or the IBM Cell and Sun Niagara, are examples of this trend. One of the main reasons that motivate designers to use CMP is the reduced design time necessary for its development, since the employed processors are usually already validated, allowing the reuse of existing designs. This way, all the effort is focused on the communication between the components.

This way, extending reconfigurable architectures, following the CMP strategy, can be a good focus of research. As an example, Fig. 7.4a shows a general overview of how a regular reconfigurable architecture could be implemented: the communication between the components of the architecture and the processor is done using dedicated buses, which makes its implementation not scalable considering the increment on the number of available RFUs. Figure 7.4b, in turn, illustrates how a CMP

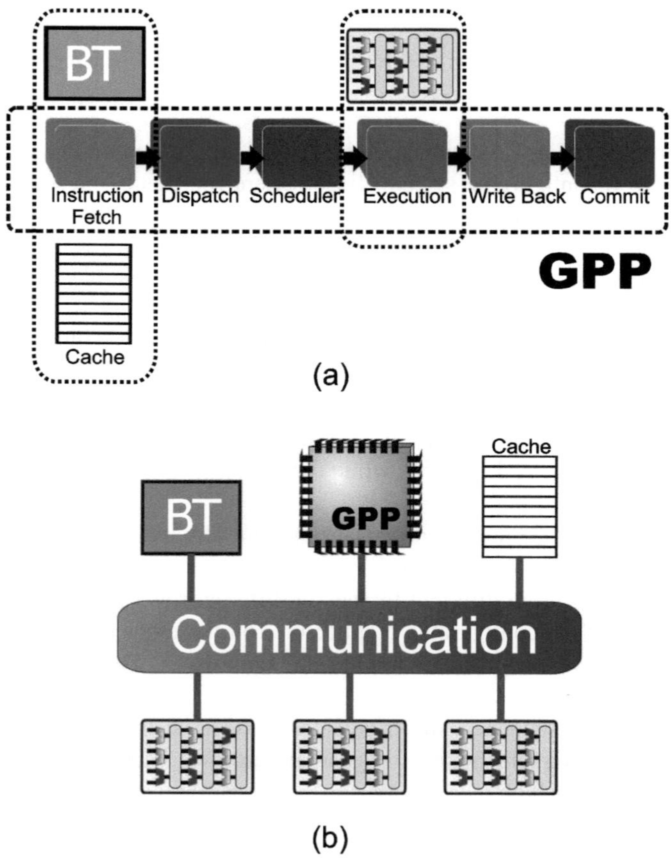

Fig. 7.4 (**a**) Usual implementation (**b**) reconfigurable architecture based on CMP

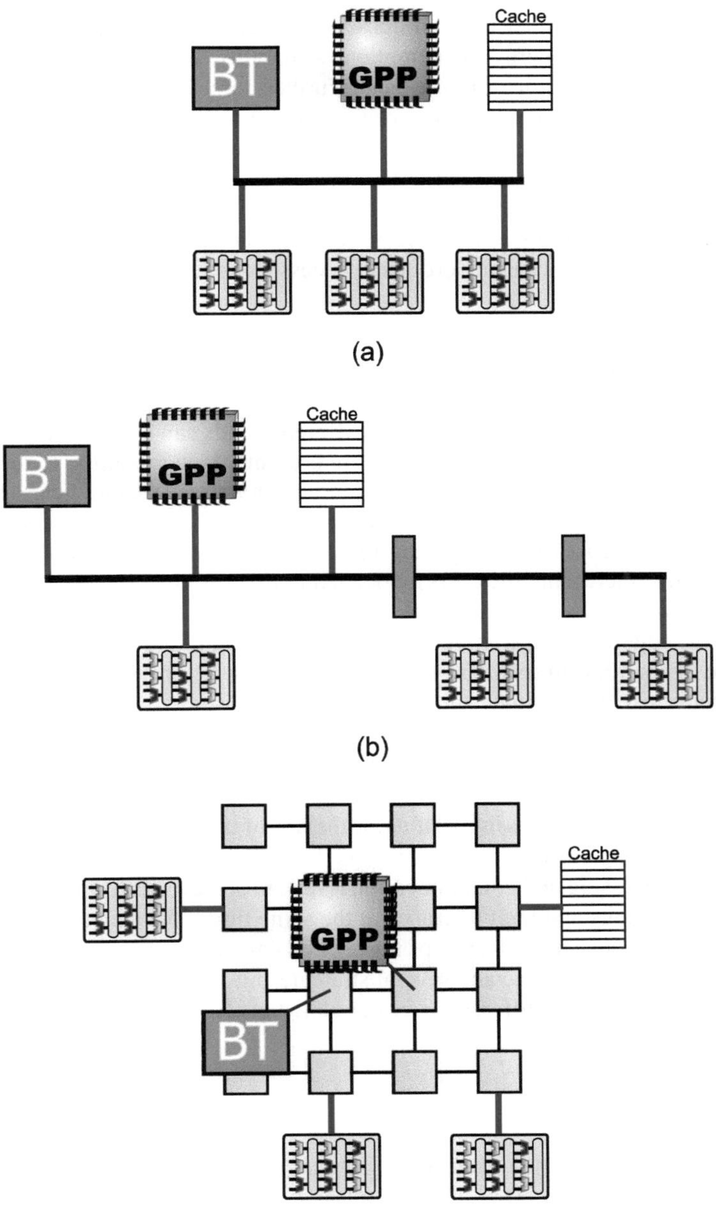

Fig. 7.5 Communication alternatives. (**a**) Monolithic bus (**b**) Segmented bus (**c**) Intra chip network

model could be implemented. With a new communication mechanism, it would be possible to increase the number of RFUs.

There is a great number of open questions concerning reconfigurable CMP architectures, such as: energy consumption, scalability, testability, fault tolerance, reusability, partitioning of processes, etc. Furthermore, it is also necessary to analyze the communication means between the components, as well as memory sharing, such as monolithic buses Fig. 7.5a or segmented (Fig. 7.5b); or the use of a crossbar or even intrachip networks (Fig. 7.5c). Finally, the possibility of implementing a heterogeneous architecture, composed of different reconfigurable units that can be used according to the process requirements at a given moment could be evaluated. Similar studies using ordinary processors were done in [193].

7.9 Final Considerations

Systems will have to change and evolve. Different trends can be observed in the embedded systems industry, for its products are presently being required to run several different applications with distinct behaviors, becoming even more heterogeneous, with extra pressure on power and energy consumption. Furthermore, while transistor size shrinks, processors are getting more sensitive to fabrication defects, aging and soft faults, increasing the costs associated to their production. To make this situation even worse, designers are stuck with the need to sustain binary compatibility, in order to support the huge amount of software already deployed.

In this scenario, different hardware resources must be provided at different levels: to better execute a single thread, according to a given set of constraints at a certain time; to allocate resources and schedule different processes depending on availability, performance requirements and the energy budget; or to sustain working conditions when a fault occurs during run time, or to increase yield to allow cost reductions even with aggressive scaling or the use of unreliable technologies.

In this changing scenario, adaptability is the key. Adaptive systems will have to work at the processing and communication levels, to achieve performance optimization, energy savings and fault tolerance at the same time. The techniques discussed throughout this book show clear steps towards this main objective. However, there is still a lot of work to be done and several strategies must be continuously developed together to achieve such different and interrelated goals.

References

182. Ahmed, E., Rose, J.: The effect of lut and cluster size on deep-submicron fpga performance and density. In: FPGA '00: Proceedings of the 2000 ACM/SIGDA Eighth International Symposium on Field Programmable Gate Arrays, pp. 3–12. ACM, New York (2000). doi:10.1145/329166.329171
183. Anderson, J.H., Najm, F.N.: Low-power programmable routing circuitry for fpgas. In: ICCAD '04: Proceedings of the 2004 IEEE/ACM International Conference on Computer-aided Design, pp. 602–609. IEEE Computer Society, Los Alamitos (2004). doi:10.1109/ICCAD.2004.1382647

184. Beck, A.C.S., Rutzig, M.B., Gaydadjiev, G., Carro, L.: Transparent reconfigurable acceleration for heterogeneous embedded applications. In: DATE '08: Proceedings of the Conference on Design, Automation and Test in Europe, pp. 1208–1213. ACM, New York (2008). doi:10.1145/1403375.1403669
185. Borkar, S.: Microarchitecture and design challenges for gigascale integration. In: MICRO 37: Proceedings of the 37th Annual IEEE/ACM International Symposium on Microarchitecture, p. 3. IEEE Computer Society, Los Alamitos (2004). doi:10.1109/MICRO.2004.24
186. Cheatham, J.A., Emmert, J.M., Baumgart, S.: A survey of fault tolerant methodologies for fpgas. ACM Trans. Des. Autom. Electron. Syst. **11**(2), 501–533 (2006). doi:10.1145/1142155.1142167
187. Davis IV, N.J., Gray, F.G., Wegner, J.A., Lawson, S.E., Murthy, V., White, T.S.: Reconfiguring fault-tolerant two-dimensional array architectures. IEEE Micro **14**(2), 60–69 (1994). doi:10.1109/40.272839
188. DeHon, A., Naeimi, H.: Seven strategies for tolerating highly defective fabrication. IEEE Des. Test **22**(4), 306–315 (2005). doi:10.1109/MDT.2005.94
189. Ferreira, R., Laure, M., Beck, A.C., Lo, T., Rutzig, M., Carro, L.: A low cost and adaptable routing network for reconfigurable systems. In: 23nd IEEE International Symposium on Parallel and Distributed Processing, IPDPS 2009, Rome, Italy, May 23–29, 2009, pp. 1–8. IEEE Press, New York (2009)
190. Ferreira, R., Laure, M., Rutzig, M.B., Beck, A.C., Carro, L.: Reducing interconnection cost in coarse-grained dynamic computing through multistage network. In: FPL 2008, International Conference on Field Programmable Logic and Applications, Heidelberg, Germany, 8–10 September 2008, pp. 47–52. IEEE Press, New York (2008)
191. Goldstein, S.C., Schmit, H., Budiu, M., Cadambi, S., Moe, M., Taylor, R.R.: Piperench: A reconfigurable architecture and compiler. Computer **33**(4), 70–77 (2000). doi:10.1109/2.839324
192. Magalhaes, M.P., Carro, L.: Automatic dataflow execution with reconfiguration and dynamic instruction merging. In: IFIP VLSI-SoC 2009, IFIP WG 10.5 International Conference on Very Large Scale Integration of System-on-Chip, Florianopolis, Brazil, 12–14 October 2009. IEEE Press, New York (2009)
193. Olukotun, K., Nayfeh, B.A., Hammond, L., Wilson, K., Chang, K.: The case for a single-chip multiprocessor. In: ASPLOS-VII: Proceedings of the Seventh International Conference on Architectural Support for Programming Languages and Operating Systems, pp. 2–11. ACM, New York (1996). doi:10.1145/237090.237140
194. Singh, H., Lee, M.H., Lu, G., Bagherzadeh, N., Kurdahi, F.J., Filho, E.M.C.: Morphosys: an integrated reconfigurable system for data-parallel and computation-intensive applications. IEEE Trans. Comput. **49**(5), 465–481 (2000). doi:10.1109/12.859540
195. Stitt, G., Vahid, F.: Thread warping: a framework for dynamic synthesis of thread accelerators. In: CODES+ISSS '07: Proceedings of the 5th IEEE/ACM International Conference on Hardware/Software Codesign and System Synthesis, pp. 93–98. ACM, New York (2007). doi:10.1145/1289816.1289841
196. Stott, E., Sedcole, N.P., Cheung, P.Y.K.: Fault tolerant methods for reliability in fpgas. In: FPL 2008, International Conference on Field Programmable Logic and Applications, Heidelberg, Germany, 8–10 September 2008, pp. 415–420. IEEE Press, New York (2008)
197. Tullsen, D.M., Eggers, S.J., Levy, H.M.: Simultaneous multithreading: maximizing on-chip parallelism. In: ISCA '98: 25 Years of the International Symposia on Computer Architecture (Selected Papers), pp. 533–544. ACM, New York (1998). doi:10.1145/285930.286011

Index

A.C. Schneider Beck Fl., L. Carro, *Dynamic Reconfigurable Architectures and Transparent Optimization Techniques*,
DOI 10.1007/978-90-481-3913-2, © Springer Science+Business Media B.V. 2010

MIX
Papier aus verantwortungsvollen Quellen
Paper from responsible sources
FSC® C105338

If you have any concerns about our products,
you can contact us on
ProductSafety@springernature.com

In case Publisher is established outside the EU,
the EU authorized representative is:
Springer Nature Customer Service Center GmbH
Europaplatz 3, 69115 Heidelberg, Germany

Printed by Libri Plureos GmbH
in Hamburg, Germany